D1602007

Handbook of Logic and Proof Techniques for Computer Science

G. Krantz

Handbook of Logic and Proof Techniques for Computer Science

With 16 Figures

BIRKHÄUSER BOSTON • SPRINGER NEW YORK

Steven G. Krantz
Department of Mathematics
Washington University
One Brookings Drive
St. Louis, MO 63130
USA

Library of Congress Cataloging-in-Publication Data
Krantz, Steven G. (Steven George), 1951–
Handbook of logic and proof techniques for computer science / Steven G. Krantz.
p. cm.
Includes bibliographical references and index.
ISBN 0-8176-4220-X (alk. paper)
1. Computers. 2. Electronic data processing. I. Title.
QA76 .K723 2002
004—dc21 2001043153
CIP

Printed on acid-free paper.
© 2002 Birkhäuser Boston

All rights reserved. This work may not be translated or copied in whole or in part without the written permission of the publisher (Birkhäuser Boston, c/o Springer-Verlag New York, Inc., 175 Fifth Avenue, New York, NY 10010, USA), except for brief excerpts in connection with reviews or scholarly analysis. Use in connection with any form of information storage and retrieval, electronic adaptation, computer software, or by similar or dissimilar methodology now known or hereafter developed is forbidden.
The use of general descriptive names, trade names, trademarks, etc., in this publication, even if the former are not especially identified, is not to be taken as a sign that such names, as understood by the Trade Marks and Merchandise Marks Act, may accordingly be used freely by anyone.

ISBN 0-8176-4220-X
ISBN 3-7643-4220-X SPIN 10850279

Production managed by Louise Farkas; manufacturing supervised by Joe Quatela.
Typeset by the author in LaTeX2e.
Printed and bound by Maple-Vail Book Manufacturing Group, York, PA.
Printed in the United States of America.

9 8 7 6 5 4 3 2 1

To little Hypatia, who doesn't seem to be very logical.

Contents

Preface

Logic is, and should be, the core subject area of modern mathematics. The blueprint for twentieth century mathematical thought, thanks to Hilbert and Bourbaki, is the axiomatic development of the subject. As a result, logic plays a central conceptual role. At the same time, mathematical logic has grown into one of the most recondite areas of mathematics. Most of modern logic is inaccessible to all but the specialist.

Yet there is a need for many mathematical scientists—not just those engaged in mathematical research—to become conversant with the key ideas of logic. *The Handbook of Mathematical Logic*, edited by Jon Barwise, is in point of fact a handbook written by logicians for other mathematicians. It was, at the time of its writing, encyclopedic, authoritative, and up-to-the-moment. But it was, and remains, a comprehensive and authoritative book for the *cognoscenti*. The encyclopedic *Handbook of Logic in Computer Science* by Abramsky, Gabbay, and Maibaum is a wonderful resource for the professional. But it is overwhelming for the casual user. There is need for a book that introduces important logic terminology and concepts to the working mathematical scientist who has only a passing acquaintance with logic. Thus the present work has a different target audience.

The intent of this handbook is to present the elements of modern logic, including many current topics, to the reader having only basic mathematical literacy. Certainly a college minor in mathematics is more than sufficient background to read this book. Specifically, courses in linear algebra, finite mathematics, and mathematical structures would be more than adequate preparation. From the computer science side, it would be good if the reader knew a programming language and had some exposure to issues of complexity and decidability. But all these prerequisites are primarily for motivation. This handbook is, to the extent possible, self-contained. It will be a compact and accessible reference.

This is not a textbook; it is a handbook. It contains very few proofs. What it does contain are definitions, examples, and discussion of all of the key ideas in basic logic. We also include cogent and self-contained introductions to critical advanced topics, including Gödel's completeness and incompleteness theorems, methods of proof, cardinal and ordinal numbers, the continuum hypothesis, the axiom of choice, model theory, number systems and their construction, multi-valued logics, category theory, universal algebra, proof theory, fuzzy set theory, recursive functions, **NP**-completeness, decision problems, Boolean algebra, semantics, decision problems, and the word problem.

This book is intended to be a resource for the working mathematical scientist. The computer scientist or engineer or system scientist who must have a quick sketch of a key idea from logic will find it here in self-contained form, accessible to a quick read. In addition to critical terminology, notation, and ideas, references for further reading are provided. There is a thorough index, a glossary, a lexicon of notation, a guide to the literature, and an extensive bibliography.

There is no other book like this one on the subject of logic and its allied areas. This book is both a reference and a portal to further study of topics in logic. A scientist reading a technical tract in another field and encountering an unfamiliar term or concept from logic can turn to this volume and find a rapid introduction to the key points. The book will be a useful reference both for working scientists and for students.

Certainly computer science is one of the most active areas of modern scientific activity that uses logic. This book has been written with computer scientists in mind. Important ideas, such as recursion theory, decidability, independence, completeness, consistency, model theory, **NP**-completeness, and axiomatics are presented here in a form that is particularly accessible to computer scientists. Examples from computer science are provided whenever possible. A special effort has been made to cut through the mathematical formalism, difficult notation, and esoteric terminology that are typical of modern mathematical logic. The treatment of any key topic is quick, incisive, and to the point.

Although this is a handbook of logic, it is not strictly logical in nature. By this we mean that it is virtually impossible to present all of the topics of this great tapestry in a strictly logical order. For that reason, topics have been repeated or presented in unusual sequence when convenient for local use.

Birkhäuser engaged a number of experts to review various versions of this project and to help prepare it for publication. These reviewers patiently guided me regarding both form and substance. Their criticisms and remarks have been invaluable. Of course, responsibility for all remaining errors and malapropisms resides entirely with the author.

It is hoped that this work will stimulate the communication between computer science and mathematics, and also be a useful handbook for both fields. It is safe to say that computer science is becoming ever more mathematical and also that mathematics is becoming increasingly comfortable with the use of computers. This book should serve as a catalyst in both activities. It is both a touchstone for a first look at topics in logic and an invitation to further reading. We hope that it will awaken in others the fascination with logic that we have experienced for more than a quarter century.

Steven G. Krantz
St. Louis, Missouri

Chapter 1

Notation and First-Order Logic

... Even though his most important research contributions were in topology, Brouwer never gave courses on topology, but always on—and only on—the foundations of intuitionism. It seemed that he was no longer convinced of his results in topology because they were not correct from the point of view of intuitionism, and he judged everything that he had done before, his greatest output, false according to his philosophy.

—B.L. Van der Waerden

Most mathematicians would find it hard to believe that there could be any serious controversy about the foundations of mathematics, any controversy whose outcome could significantly affect their own mathematical activity.

—Errett Bishop

The world is all that is the case.

—Ludwig Wittgenstein

When Aristotle was asked how much educated men were superior to the uneducated, he replied, "As much as the living are to the dead."

—Dionysus of Halicarnassus

The world is a totality of facts, not things.

—Ludwig Wittgenstein

Mathematics takes us into the region of absolute necessity, to which not only the actual world, but every possible world, must conform.

—Bertrand Russell

If the only tool you have is a hammer, then every problem you see looks like a nail.

—Confucian saying

1.1 The Use of Connectives

1.1.1 Elementary Statements

An *elementary statement* (or *atomic statement*) is a sentence with a subject and a verb (and sometimes an object) but no connectives (and, or, not, if-then, if-and-only-if). See Subsection 2.1.1 for a more rigorous definition. Elementary statements are joined together using connectives. Thus are formed *well-formed formulas*, or *wffs*.

Examples of elementary statements are:

George is good.

Steve has hair.

John is tall.

In the propositional calculus, a *sentence* is composed of *elementary statements* joined together with *connectives.*

1.1.2 Connectives

The basic sentential connectives are "and," "or," "not," "if-then," and "if-and-only-if." Thus, we may join two elementary statements together using a connective:

[George is good] and [John is tall].

In ordinary parlance, we do not include the brackets [ ], but we use them in this instance for clarity.

In practice, it is best to use symbols. We denote an elementary statement by a capital boldface roman letter: $\mathbf{A}, \mathbf{B}, \mathbf{E}, \mathbf{G}, \mathbf{H}$. The connectives are denoted as follows:

and	$\wedge$
or	$\vee$
not	$\sim$
if-then	$\Rightarrow$
if-and-only-if	$\Longleftrightarrow$

Thus, if

$$\mathbf{A} = \text{“}\mathbf{John\ is\ tall.}\text{”}$$

and

$$\mathbf{E} = \text{“}\mathbf{George\ is\ good.}\text{”}$$

then

$$\mathbf{A} \vee \mathbf{E}$$
$$\mathbf{E} \Rightarrow \mathbf{A}$$
$$\mathbf{A} \wedge \mathbf{E}$$
$$\sim \mathbf{E}$$
$$\mathbf{E} \iff \mathbf{A}$$

denote, respectively,

[John is tall] or [George is good].
If [George is good], then [John is tall].
[John is tall] and [George is good].
It is not the case that [George is good].
[George is good] if and only if [John is tall].

We often summarize the penultimate statement by saying "George is not good." The very last statement can be alternatively phrased as "Necessary and sufficient for George to be good is that John be tall." The statement means that both "If George is good, then John is tall" and "If John is tall, then George is good."

We sometimes refer to the "and" operation as "conjunction" and to the "or" operation as "disjunction."

A final note is that there are other notations for the logical connectives. Some books use $\supset$ instead of $\Rightarrow$. Others use $\neg$ instead of $\sim$. Other variations may be found in the literature; these are usually clear from context. The classic [SUP1] is a good source for all matters discussed here.

1.1.3 *Redundancy of the Connectives*

The commonly used connectives are redundant. For example, $\mathbf{A} \Rightarrow \mathbf{B}$ is logically equivalent to $\sim \mathbf{A} \vee \mathbf{B}$. And $\mathbf{A} \wedge \mathbf{B}$ is logically equivalent to $\sim (\sim \mathbf{A} \vee \sim \mathbf{B})$. In fact, all of the propositional calculus can be formulated using only $\sim, \vee$. Alternatively, all of the propositional calculus can be formulated using only $\sim, \wedge$. A third alternative is to use only $\sim, \Rightarrow$. It is *not* possible to formulate the propositional calculus using only $\vee, \wedge$ or $\vee, \Rightarrow$ or $\wedge, \Rightarrow$. Truth tables (see the next section) can be used to provide proofs of all of these statements.

1.1.4 *Additional Connectives*

In the classical theory of electrical circuits (whose theoretical realization is Boolean algebra—see Chapter 13), it is convenient to use two additional connectives. These are "nand" (denoted $\mid$) and "nor" (denoted $\downarrow$). The first of these denotes the denial of conjunction: $\mathbf{A} \mid \mathbf{B} \equiv \sim (\mathbf{A} \wedge \mathbf{B})$. The second denotes the denial of disjunction:

$\mathbf{A} \downarrow \mathbf{B} \equiv \sim (\mathbf{A} \vee \mathbf{B})$. Part of the interest of "nand" and "nor" is that either one of these connectives, *by itself*, can be used to express all statements in the propositional calculus. For instance, $\mathbf{A} \downarrow \mathbf{A}$ is logically equivalent to $\sim \mathbf{A}$ and $[\mathbf{A} \downarrow \mathbf{A}] \downarrow [\mathbf{B} \downarrow \mathbf{B}]$ is logically equivalent to $\mathbf{A} \wedge \mathbf{B}$. Also, we know that $\sim$ and $\wedge$ suffice to generate $\vee$ and $\Rightarrow$ and hence all statements of the propositional calculus. A similar discussion applies to "nand." Although "nor" has the elegant "primitive universal" property just described, it is *not* associative. Therefore, it is not as useful as the more familiar connectives.

Although not often used in mathematics, there is another version of "or" besides the one that we use in this book. The mathematical "**A** or **B**," as used here, means "either **A** or **B** or both." This is sometimes called the "inclusive 'or'," and we write it $\mathbf{A} \vee \mathbf{B}$. The other, nonmathematical "or" is the "exclusive 'or'." It is written $\mathbf{A} \oplus \mathbf{B}$, and it means "**A** or **B** but not both." We sometimes refer to this connective as *xor*.

1.2 Truth Values and Truth Tables

1.2.1 Rules for Truth Values and Tables

In logic, every elementary statement has a truth value. Usually, these truth values are called "true" and "false" and are denoted **T** and **F**. (There is a theory of multivalued logics in which there are more than two truth values. We shall not explore that subject here, but see Subsection 1.2.2.) When we form compound sentences using the connectives, the truth values of the new sentences can then be computed. The basic rules are these:

A	B	$\mathbf{A} \wedge \mathbf{B}$
T	**T**	**T**
T	**F**	**F**
F	**T**	**F**
F	**F**	**F**

A	B	$\mathbf{A} \vee \mathbf{B}$
T	**T**	**T**
T	**F**	**T**
F	**T**	**T**
F	**F**	**F**

A	$\sim$ A
T	F
F	T

A	B	A $\Rightarrow$ B
T	T	T
T	F	F
F	T	T
F	F	T

A	B	A $\iff$ B
T	T	T
T	F	F
F	T	F
F	F	T

These displays are called *truth tables*. Each of them makes good intuitive sense. Consider, for example, the first one. The statement "**A** and **B**" means that *both* **A** and **B** are true. Thus, the compound statement can be true only if each of the component elementary statements is true. That is what the truth table says. A similar type of analysis may be applied to each of the other tables.

1.2.2 Multivalued Logics

There exist theories of multivalued logics. In a multivalued logic, there are not just two truth values (**T** and **F**) but many truth values. The most-studied multivalued logics are three-valued, and a prime example is *modal logic* (Subsection 1.2.3). The reader is referred to [STA] or [IYK] for a detailed treatment of modal logic. Suffice it to say that this logic, useful both to philosophers and computer scientists, allows for statements that are "necessarily true," "possibly true," "known to be true," "true in the future," or "believed to be true." As a simple example, the statement

Bill Clinton is President of the United States

used to be true but is no longer true. The statement

George W. Bush is President of the United States

is true in 2001; it was not prior to 2001, and will not be true after 2008. The sentence

There are nine planets in the solar system

is true as best we know. It describes an observed fact that is correct right now. It may not always have been correct, and it is not *necessarily correct* (in the sense that $2 \times 2 = 4$ is necessarily correct).

The third truth value allows for this new notion of indefinite truth. Computer scientists are interested in modal logic because of its intersection with the theory of fuzzy sets and fuzzy logic—see Section 5.10 and, for example, [PIA]. Variants of modal logic include deontic logic, temporal logic, and doxastic logic.

1.2.3 Modal Logic

The most common formulation of modal logic is a weak logic called K, in honor of Saul Kripke. This logic includes the usual connectives as well as a "modal operator" $\Box$. The modal operator means "it is necessary that." Other operators of interest can be defined from the modal operator. For example, an operator $\diamond$, meaning "possibly true," can be defined as

$$\diamond A = \sim \Box \sim A.$$

Modal logic necessitates new rules for the use of the quantifiers $\exists$ and $\forall$. We cannot indulge in the details here but refer the interested reader to [STA].

1.2.4 Compound Sentences and Truth Values

A compound sentence may be analyzed for its truth value by reduction to its elementary parts.

Example 1.1

Consider the statement $(\mathbf{A} \vee \mathbf{B}) \Rightarrow (\sim \mathbf{A} \wedge \mathbf{B})$. We analyze its truth values as follows:

A	B	$\mathbf{A} \vee \mathbf{B}$	$\sim \mathbf{A} \wedge \mathbf{B}$	$(\mathbf{A} \vee \mathbf{B}) \Rightarrow (\sim \mathbf{A} \wedge \mathbf{B})$
T	**T**	**T**	**F**	**F**
T	**F**	**T**	**F**	**F**
F	**T**	**T**	**T**	**T**
F	**F**	**F**	**F**	**T**

□

Two statements are said to be *logically equivalent* if they have the same truth table. For example,

$$\mathbf{A} \vee (\mathbf{B} \wedge \sim \mathbf{A})$$

and

$$\sim (\sim \mathbf{A} \wedge (\sim \mathbf{B} \vee \mathbf{A}))$$

are logically equivalent. We see this by checking the truth tables:

A	B	$\sim$ A	B$\wedge \sim$ A	A $\vee$ (B$\wedge \sim$ A)
T	**T**	**F**	**F**	**T**
T	**F**	**F**	**F**	**T**
F	**T**	**T**	**T**	**T**
F	**F**	**T**	**F**	**F**

A	B	$\sim$ B $\vee$ A	$\sim$ A $\wedge$ ($\sim$ B $\vee$ A)	$\sim$ ($\sim$ A $\wedge$ ($\sim$ B $\vee$ A))
T	**T**	**T**	**F**	**T**
T	**F**	**T**	**F**	**T**
F	**T**	**F**	**F**	**T**
F	**F**	**T**	**T**	**F**

There are a number of elementary relationships among the connectives. We enunciate a few of these here

(1) $[\mathbf{A} \Rightarrow \mathbf{B}] \Leftrightarrow [\sim \mathbf{A} \vee \mathbf{B}]$

(2) $[\mathbf{A} \Rightarrow \mathbf{B}] \Leftrightarrow [\sim (\mathbf{A} \wedge \sim B)]$

(3) (de Morgan's Law) $\sim (\mathbf{A} \vee \mathbf{B}) \Leftrightarrow \sim A \wedge \sim \mathbf{B}$

(4) (de Morgan's Law) $\sim (\mathbf{A} \wedge \mathbf{B}) \Leftrightarrow \sim A \vee \sim \mathbf{B}$

1.2.5 Tautologies and Contradictions

If a statement φ is always true, no matter what the truth values of its components, then the statement is called a *tautology*. We will sometimes write $\models \varphi$. If a statement is false, no matter what the truth values of its components, then the statement is called a *contradiction*. The reader may check that

$$[\mathbf{A} \wedge (\sim \mathbf{A} \vee \mathbf{B})] \Rightarrow \mathbf{B}$$

is a tautology, whereas

$$\mathbf{A} \wedge \sim \mathbf{A}$$

is a contradiction.

1.2.6 Contrapositives

The statement $\sim \mathbf{B} \Rightarrow \sim \mathbf{A}$ is logically equivalent with the statement $\mathbf{A} \Rightarrow \mathbf{B}$, as can be checked with a truth table. The first statement is called the *contrapositive* of the second.

The contrapositive should be carefully distinguished from the *converse*. The converse of the statement $\mathbf{A} \Rightarrow \mathbf{B}$ is $\mathbf{B} \Rightarrow \mathbf{A}$. The truth or falsity of the converse of a statement is independent of the truth or falsity of the original statement. For example,

$$x > 0 \Rightarrow x^2 > 0$$

is true while its converse is false. On the other hand,

$$x > 2 \Rightarrow x + 5 > 7$$

is true and its converse is true as well.

1.3 The Use of Quantifiers

1.3.1 "For All" and "There Exists"

The two fundamental quantifiers are "for all" (denoted $\forall$) and "there exists" (denoted $\exists$). The symbolic statement

$$\forall x, A(x)$$

means that $A(x)$ is true for every value of x (in the domain of A). The symbolic statement

$$\exists y, B(y)$$

means that $B(y)$ is true for some (i.e., for at least one) value of y in the domain of B.

If $A(x)$ denotes the statement "The number x^2 is nonnegative," then

$$\forall x, A(x)$$

means "For every x, the number x^2 is nonnegative." This assertion is true provided that it is understood that x ranges over the real numbers.

If $B(y)$ denotes the statement "The number $y + 3$ is negative," then

$$\exists y, B(y)$$

means "There is a y such that $y + 3$ is negative." This assertion is true provided that it is understood that y ranges over the integers.

1.3.2 Relations Between "For All" and "There Exists"

The two fundamental quantifiers are related in the following important way:

$\sim \forall \sim$ is equivalent to $\exists$,
$\sim \exists \sim$ is equivalent to $\forall$.

For example, the statement

> It is not the case that, for every integer, that integer is not prime.

means just the same thing as

> There exists an integer that is prime.

Also

> It is not the case that there exists a positive real number that is not a square

means just the same thing as

> Every positive real number is a square.

The books [HAL1], [HAL2] discuss further algebraic aspects of this relationship between $\forall$ and $\exists$.

1.3.3 The Propositional and the Predicate Calculus

The rules for forming valid statements using $\wedge$, $\vee$, $\sim$, $\Rightarrow$, $\iff$ are known as the propositional calculus. It is possible to formalize the rules of the propositional calculus. In what follows, we use the term *valid* to refer to a statement that is true in every interpretation of the theory being considered.[1] We use the phrase *identically true* to refer to a statement that is true no matter what the truth values of its component variable letters. We use the phrase *propositional function* to refer to a combination of variable letters using the aforementioned connectives. The first two axioms that we list are for the propositional calculus. The subsequent five axioms augment the propositional calculus to the predicate calculus (i.e., the logical calculus that involves quantifiers).

---

[1] *Conjunctive normal form* (CNF) for a statement is particularly useful in verifying validity. A statement is in CNF if it can be written as $\mathbf{A}_1 \wedge \mathbf{A}_2 \wedge \cdots \wedge \mathbf{A}_k$, where each $\mathbf{A}_j$ is either an atomic statement or a disjunction of atomic statements. See [HUR. pp. 72–76]. This reference by Huth and Ryan has associated with it a very helpful Web site: `http://www.cs.bham.ac.uk/research/lics/`. Both the book and the Web site explore issues of how to verify computer systems and software using methods of logic. That topic is important, but it exceeds the scope of the present book. It is sometimes also useful to consider *disjunctive normal form*.

A (Rule of the Propositional Calculus) If **P** denotes a propositional function of the variable letters $A_1, \ldots, A_n$ that is identically true, then the result of replacing each A_j by any other statement is a valid statement.

B (Rule of Inference—*Modus Ponens*) If A and $A \Rightarrow B$ are valid statements, then so is B.

The next statements are part of the predicate calculus.

C (Rule of Equality) If c, c', c'' are constant symbols, then the statements $c = c$, $[c = c'] \Rightarrow [c' = c]$, and $\big([c = c'] \wedge [c' = c'']\big) \Rightarrow [c = c'']$ are valid statements.

If A is a statement, if c and c' are constant symbols, and if A' represents A with every occurrence of c replaced by c', then $[c = c'] \Rightarrow \big(A \Rightarrow A'\big)$ is a valid statement.

D (Change of Variables) If A is any statement and A' results from A by replacing each occurrence of the symbol x with the symbol x', where x and x' are any two variable symbols, then the statement $A \iff A'$ is a valid statement.

E (Rule of Specialization) If c is any constant symbol, then the statement $\forall x, A(x) \Rightarrow A(c)$ is a valid statement.

F (Axiom for Existence) Let B be any statement not involving the constant c or the variable x. If $A(c) \Rightarrow B$ is valid, then so is $\exists x, A(x) \Rightarrow B$.

G (Axiom for Prenex Normal Form) Let $A(x)$ have x as the only free variable, and let every occurrence of x be free (see Subsection 1.5.1 for a detailed treatment of the concept of free variable). Let B be a statement that does not contain x. Then the following are valid statements:

$$\begin{aligned}
[\sim \forall x, A(x)] &\iff [\exists x, \sim A(x)], \\
[\forall x, (A(x) \wedge B)] &\iff [(\forall x, A(x)) \wedge B)], \\
[\exists x, (A(x) \wedge B)] &\iff [(\exists x, A(x)) \wedge B)].
\end{aligned}$$

1.3.4 Derivability

Definition 1.1 Let S be a collection of statements. The statement A is *derivable* from S if, for some $B_1, \ldots, B_n \in S$, the statement $[B_1 \wedge \cdots \wedge B_n] \Rightarrow A$ is true.

1.3.5 Semantics and Syntax

In formal logic, a notion that involves interpretation (truth and models) is called *semantical.* By contrast simple relations among symbols and expressions of precise formal languages are called *syntactical.* Because of the many paradoxes in set theory, it is usually considered safest to stick to the syntactical study of foundations. Much of the point of modern logic is to sort out the semantical from the syntactical.

1.3.6 A Consideration of First-Order Theories

We conclude this section with some common logical terminology.

An *n-ary function* from a set A to a set B is a function

$$f : \underbrace{A \times A \times \cdots \times A}_{n \text{ times}} \to B.$$

A subset of the n-fold product $A \times A \times \cdots \times A$ is called an *n-ary predicate.* If $\mathbf{P}$ is such a predicate, then the notation $P(a_1, \ldots, a_n)$ is sometimes used to mean that the n-tuple $(a_1, \ldots, a_n)$ is in $\mathbf{P}$.

Definition 1.2 A *first-order language* has the following components:

- as variables,

$$x, y, z, w, z', y', z', w', x,'' y,'' z,'' w,'' \ldots;$$

- for each n, the n-ary function symbols and the n-ary predicate symbols;
- the symbols $\sim$, $\vee$, and $\exists$.

(Again, bear in mind that $\wedge$, $\Rightarrow$, and $\iff$ can be expressed in terms of $\sim$ and $\vee$. Also, $\forall$ can be expressed in terms of $\sim$ and $\exists$.)

We now give a syntactical definition of first-order logic.

Definition 1.3 A *first-order theory* T has the following properties.

1. The language of T is a first-order language.
2. The axioms of T are the logical axioms and certain further non-logical axioms.
3. The rules of inference in T are:

   Expansion Rule Infer $\mathbf{B} \vee \mathbf{A}$ from $\mathbf{A}$.

Contraction Rule Infer **A** from **A** $\vee$ **A**.

Associative Rule Infer (**A** $\vee$ **B**) $\vee$ **C** from **A** $\vee$ (**B** $\vee$ **C**).

Cut Rule Infer **B** $\vee$ **C** from **A** $\vee$ **B** and $\sim$ **A** $\vee$ **C**.

$\exists$-Introduction Rule If x is not free in **B**, infer $\exists \boldsymbol{x}, \mathbf{A} \to \mathbf{B}$.

It is common to say that *first-order logic* consists of the connectives $\wedge$, $\vee$, $\sim$, $\Rightarrow$, $\iff$, the equality symbol $=$, and the quantifiers $\forall$ and $\exists$, together with an infinite string of variables $x, y, z, \ldots, x', y', z', \ldots$ and, finally, parentheses (,) to keep things readable (see [BAR, p. 7]). The word "first" here is used to distinguish the discussion from second-order and higher-order logics. In first-order logic, the quantifiers $\forall$ and $\exists$ always range over elements of the domain M of discourse. Second-order logic, by contrast, allows us to quantify over subsets of M and functions F mapping $M \times M$ into M. Third-order logic treats sets of function and more abstract constructs. The distinction among these different orders is often moot.

First-order logic is a substantial extension of propositional logic. It allows reasoning about individuals using functions and predicates acting on individuals. Many mathematicians (and even logicians!) would say that all mathematics can be formulated in first-order logic. This latter point of view, formulated a bit more precisely, is commonly referred to as *Hilbert's thesis* (see [BAR, p. 41]).

1.3.7 Herbrand's Theorem

Herbrand's theorem is a milestone in proof theory. It gives sufficient conditions for reduction of first-order logic to propositional logic. In the statement of this result, we use the phrase "universal formula" to mean a formula of the form $\forall x_1, \ldots \forall x_k, \theta$, where θ is quantifier-free.

Theorem 1.1 (Herbrand)
Let T be a theory axiomatized by universal formulas. Suppose that $T \models \forall \mathbf{x}, \exists y_1, \ldots \exists y_k, B(\mathbf{x}, \mathbf{y})$, with $B(\mathbf{x}, \mathbf{y})$ a quantifier-free formula and $\mathbf{x} = (x_1, \ldots, x_n)$, $\mathbf{y} = (y_1, \ldots, y_k)$. Then there is a finite sequence of terms $t_{ij} = t_{ij}(\mathbf{x})$ with $1 \le i \le r$ and $1 \le j, k$ such that

$$T \vdash \forall \mathbf{x}, \left(\bigvee_{i=1}^{r} B(\mathbf{x}, t_{i1}, \ldots, t_{ik}) \right).$$

(Here the symbol $\bigvee$ is a disjunction of sets indexed by $1, \ldots, r$. The notation $T \models \psi$ means that any model that makes all the statements of T true also makes ψ true. Finally, $T \vdash \psi$ means that ψ is provable from T.)

As we see, Herbrand's theorem gives a paradigm for eliminating the quantifier on the variable $\mathbf{y}$. Of course it may be applied iteratively to eliminate several quantifiers.

1.3.8 An Example from Group Theory

To clarify this last point, consider the following definition. An abelian group G is said to be *divisible* if

$$\forall n \geq 1, \forall x, \exists y, [ny = x]. \qquad (*)$$

This statement is *not* a first-order statement because the leading quantifier ranges over the set of natural numbers (rather than over the domain of discourse G). But one may sidestep the issue (as does Cohen in [COH]) by using the following infinite list of axioms:

$$\begin{aligned} &\forall x, \exists y, [2y = x]; \\ &\forall x, \exists y, [3y = x]; \\ &\forall x, \exists y, [4y = x]; \\ &\qquad\qquad\quad \cdots \\ &\forall x, \exists y, [ny = x]; \\ &\qquad\qquad\quad \cdots . \end{aligned}$$

(Observe that we omit the initial statement $\forall x, \exists y, [y = x]$ because it is trivial.) Thus the second-order statement $(*)$ can be replaced by infinitely many first-order statements. For the sake of the discussion in the present handbook, the distinction between first-order and second-order is insignificant, and we will rarely make further reference to it.

1.4 Gödel's Completeness Theorem

1.4.1 Provable Statements and Tautologies

Gödel's completeness theorem has both an abstract formulation and a concrete formulation. We provide both. The theorem, in its most concrete rendition, says in effect that any true statement that can be formulated in the propositional calculus can be proved in the propositional calculus. This means that there is a finite sequence of valid logical steps that culminates in the desired statement. Before enunciating the theorem, let us consider an example.

Example 1.2

Consider the statement $\mathbf{B} \vee \sim \mathbf{B}$. This statement is a tautology. It is true all the time, as the truth table shows:

$\mathbf{B}$	$\sim \mathbf{B}$	$\mathbf{B} \vee \sim \mathbf{B}$
T	**F**	**T**
F	**T**	**T**

□

Gödel's theorem asserts that, because this statement is universally true, it has a proof. Here is the proof (using the syllogisms from Subsection 1.3.6):

Proof of $\mathbf{B} \vee \sim B$

(1)	$(\mathbf{B} \vee \mathbf{B}) \Rightarrow \mathbf{B}$	by the Contraction Rule
(2)	$\sim (\mathbf{B} \vee \mathbf{B}) \vee \mathbf{B}$	by truth table from (1)
(3)	$B \Rightarrow (\mathbf{B} \vee \mathbf{B})$	by the Expansion Rule
(4)	$\sim \mathbf{B} \vee (\mathbf{B} \vee \mathbf{B})$	by truth table from (3)
(5)	$(\mathbf{B} \vee \mathbf{B}) \vee \sim \mathbf{B}$	by truth table from (4)
(6)	$(\sim (\mathbf{B} \vee \mathbf{B}) \vee \mathbf{B}) \wedge ((\mathbf{B} \vee \mathbf{B}) \vee \sim \mathbf{B}))$	conjunction of (2) and (5)
(7)	$[(\sim (\mathbf{B} \vee \mathbf{B}) \vee \mathbf{B}) \wedge ((\mathbf{B} \vee \mathbf{B}) \vee \sim \mathbf{B})] \Rightarrow \mathbf{B} \vee \sim \mathbf{B}$	by the Cut Rule
(8)	$\mathbf{B} \vee \sim \mathbf{B}$	*modus ponendo ponens* applied to (6), (7)

□

1.4.2 *Formulation of Gödel's Completeness Theorem*

Gödel's theorem says that the phenomenon illustrated in the last subsection is valid all the time. In fact it is true not just in the propositional calculus but in the predicate calculus as well. A quick and dirty formulation of Gödel's theorem is this:

> **Theorem**: Let S be a collection of statements in the predicate calculus that is consistent (i.e., it has no internal contradictions). Then there is a model for S.

An alternative (logically equivalent) formulation is

> **Theorem**: In any first-order predicate calculus, the theorems are precisely the logically valid wffs ([MEN, p. 68]).

1.4.3 Additional Terminology

Some additional, and more standard, terminology is that a collection of statements S is *consistent* if the statement $A \wedge \sim A$ cannot be derived from S for any A. A statement ϕ is *consistent* with S if one cannot prove $\sim \phi$ from S. Alternatively, ϕ is consistent with S if $S + \{\phi\}$ does not lead to any contradictions. The most convenient way to prove consistency is to find another sentence ψ that is known to be consistent with S and then to prove that $S + \{\psi\}$ implies ϕ. The hard way to prove consistency is to build a model (see Sections 3.2 and 2.4.2) in which both S and ϕ are true.

A *model* for a collection S of statements is a set M together with an interpretation of some constant and relation symbols in S. This "interpretation" consists of a map $c_\alpha \to \overline{c}_\alpha$ from the constant elements of S to elements of M and a second map $R_\beta \to \overline{R}_\beta$ from the relation elements of S to elements of M such that all the statements of S are true in M.

In practice, we often say that a collection S of statements is *consistent* (or *satisfiable*) if it has at least one model. Horn formulas and Horn clauses are very useful in checking satisfiability, as it is easier to decide the question when a statement is put in Horn's normalized form (see [HUR, pp. 85–86]).

1.4.4 Some More Formal Language

To state Gödel's result more formally, we introduce a convenient piece of terminology that will make various model-theoretic results convenient to state. (Refer to Section 3.2 for a detailed treatment of models. For now, think of a model of a theory as an interpretation of that theory.) Let us call a model M for a set S of statements *coordinated* if either of the following assertions holds (see Section 5.8 for terminology).

1. If S is finite, then M is at most countable.

2. If S is infinite, then the cardinality of M does not exceed the cardinality of S.

1.4.5 Other Formulations of Gödel Completeness

Theorem 1.2 (Gödel)
Let S be any consistent set of statements in first-order logic. Then there exists a coordinated model for S.

Theorem 1.3 (Gödel completeness for the propositional calculus)
If S is a collection of statements that contains no quantifiers and is consistent, then there is a (coordinated) model M for S in which every element of M is the image of some object in S.

To round out the picture—that is, to tie up the intuitive statement of Gödel's theorem with the more rigorous statements just presented—we formulate the following corollary of Gödel's ideas.

Proposition 1.1
If A is not derivable from S, then there is a model for S in which A is false.

Corollary 1.1
If a statement is true in every model, then it is provable.

1.4.6 The Compactness Theorem

Another nonobvious corollary of Gödel's ideas is the standard compactness theorem of sentential logic:

Theorem 1.4
In first-order logic, if every finite subset of a system S has a model, then S has a model.

Both the completeness theorem and the compactness theorem were proved in Gödel's Ph.D. thesis in 1930.

1.4.7 Tautological Implication and Provability

We close this discussion by introducing some common and very useful notation. Let T be a (finite) collection of sentences and let ψ be a statement. We write $T \models \psi$ if ψ is true in all models that make the sentences in T true (i.e., ψ is a logical consequence of T). We write $T \vdash \psi$ if ψ is provable from T. In other words, $\vdash$ is syntactic and $\models$ is semantic. The Gödel completeness theorem may now be formulated as follows.

Theorem 1.5
Let T be a (finite) collection of sentences in first order-logic, and let ψ be a statement. Then $T \vdash \psi$ if and only if $T \models \psi$.

1.5 Second-Order Logic

This handbook is primarily about first-order logic, but we will record here a few facts about second-order logic.

1.5.1 Semantics

In addition to the usual semantics of first-order logic, our second-order logic will include:

Predicate Variables For each positive integer n, we have the n-place predicate variables

$$X_1^n, X_2^n, \ldots.$$

Function Variables For each positive integer n, we have the n-place function variables

$$F_1^n, F_2^n, \ldots.$$

The usual variables (from first-order language) $\nu_1, \nu_2, \ldots$ will now be called *individual variables.* The set of valid terms is defined to be the set of expressions that can be built up from the constant symbols and the individual variables by applying the function symbols (both function parameters and function variables). Atomic formulas are, as before, expressions $Pt_1 \cdots t_n$ (or $P(t_1, \ldots, t_n)$), with $t_1, \ldots, t_n$ terms and $\mathbf{P}$ an n-place predicate symbol. A wff can be any of the valid wff's from first-order logic and in addition any formula obtained from the following formula-building operation: If φ is a wff, then so are $\forall X_i^n, \varphi$ and $\forall F_i^n, \varphi$. The notion of a free variable is the same as before (see also Section 2.3). A sentence is a wff in which no variable occurs free.

In the next example, a well-ordering is an ordering in which each set has a least element (see Subsection 5.8.10).

Example 1.3

A well-ordering can be translated into the second-order sentence

$$\forall X, [\exists y, Xy \Rightarrow \exists y, (Xy \wedge \forall z, (Xz \Rightarrow y \leq z))].$$

Here of course X followed by a variable denotes X applied to that variable.

□

Example 1.4

The induction postulate in Peano's arithmetic (Subsection 7.1.3) states that any set of natural numbers that contains 0 and is closed under the successor function is actually the set of all natural numbers. This statement can be translated into the second-order language for number theory as follows:

$$\forall X, [X0 \land \forall y, (Xy \Rightarrow XSy) \Rightarrow \forall y, Xy].$$

Here S denotes the successor function.

□

Example 1.5

The fundamental property of the ordered field of reals is that any nonempty set with an upper bound has a least upper bound. In second-order language, this says

$$\begin{aligned}&\forall X, [\exists y, \forall z, (Xz \Rightarrow z < y) \land \exists z, Xz \\ &\Rightarrow \exists y, \forall y', (\forall z, (Xz \Rightarrow z < y') \iff y \leq y')].\end{aligned}$$

□

It is known that the standard compactness theorem of first-order logic (Subsection 1.4.6) fails in second-order logic. In fact there is a set of second-order sentences that is unsatisfiable, yet every finite subset of the set is satisfiable.

Chapter 2

Semantics and Syntax

"Contrariwise," continued Tweedledee, "if it was so, it might be; and if it were so, it would be: but as it isn't, it ain't. That's logic.

—Lewis Carroll

Logic is nothing more than a knowledge of words.

—Charles Lamb

Mathematics may be defined as the subject in which we never know what we are talking about, nor whether what we are saying is true.

—Bertrand Russell

The limits of my language means the limits of my world.

—Ludwig Wittgenstein

Logic is the armory of reason, furnished with all offensive and defensive weapons.

—Thomas Fuller

Logic is an instrument used for bolstering a prejudice.

—Elbert Hubbard

O Logic: Born gatekeeper to the Temple of Science,
Victim of capricious destiny,
Doomed hitherto to be drudge of pedants,
Come to the aid of thy master, Legislation

—Jeremy Bentham

In a symbol there is concealment and yet revelation; here therefore, by Silence and by Speech acting together, comes a double significance.

—Thomas Carlyle

If you want to understand function, study structure.

—Francis Harry Compton Crick

A significant part of modern formal logic concerns semantics and syntax. Because the consideration of semantics and syntax is very closely allied with questions of computer syntax, it is important that we give a careful, if brief, treatment of some of these ideas here.

For the convenience of the reader, we again recall the basic difference between the two key terms. In formal logic, a notion that involves interpretation (truth and models) is called *semantical.* By contrast, simple relations among symbols, and expressions of precise formal languages, are called *syntactical.*

2.1 Elementary Symbols

2.1.1 Formal Systems (Syntax)

A *formal system* is a finite collection of symbols and precise rules for manipulating these symbols to form certain combinations called "theorems." The rules should be quite explicit and require no infinite processes to check. In this book we are primarily interested in mathematical systems. According to modern standards, any mathematical system will certainly contain these symbols:

$\sim$	$\wedge$	$\vee$	$\Rightarrow$	$\Longleftrightarrow$
not	and	or	implies	if and only if
$\forall$	$\exists$	$=$	$(\, , \,)$	$x, \, '$
for all	there exists	equals	parentheses	variable symbols

The symbols in the first line compose the propositional calculus. Those in the second line expand the language to first-order logic.

Any nontrivial formal language will also contain function symbols and relation symbols. A function f or a relation R will have k arguments, where k is a positive integer. If $k = 2$, then the function/relation will be called "binary." If $k = 3$, then it is called "ternary." Otherwise, we will refer to a k-ary function/relation. A 0-ary function is a *constant symbol.* (Note also that in Section 5.4 on set theory we define a notion of "relation." It is essentially the same as the use of the term here, but the context is a bit different.) Notice that $=$ is a distinguished 2-ary relation.

In Subsection 1.1.1 we gave an informal definition of elementary or atomic statements. A more rigorous definition is that an atomic statement is of the form $P(t_1, \ldots, t_k)$, for $\mathbf{P}$ a k-ary relation. In general, formulas (wffs—see the next section) are built up from atomic formulas with logical connectives.

2.2 Well-Formed Formulas or wffs [Syntax]

Next we give precise rules for forming "grammatically correct" statements. These statements are called *well-formed formulas* or wffs.

1. The statements $x = y$, $x = c$, $c = c'$ are wffs, where x and y are variables and c, c' are constants.

2. If R is a k-ary relation and if each of $t_1, \ldots, t_n$ is either a variable or a constant, then $R(t_1, \ldots, t_n)$ is a wff.

3. If $\mathbf{A}$ and $\mathbf{B}$ are wffs, then so are $\mathbf{A} \wedge \mathbf{B}$, $\mathbf{A} \vee \mathbf{B}$, $\sim \mathbf{A}$, $\mathbf{A} \Rightarrow \mathbf{B}$, and $\mathbf{A} \iff \mathbf{B}$.

4. If $\mathbf{A}$ is a wff, then so are $[\exists x, \mathbf{A}]$ and $[\forall x, \mathbf{A}]$.

2.3 Free and Bound Variables (Syntax)

Each of the variables that occurs in a wff is classified as either free or bound. From the point of view of syntax, it is essential to distinguish these two types of variables, as this difference governs what type of substitutions can be made.

(i) Every variable occurring in a formula of the form $x = y$, $x = c$, $c = c'$ or $R(t_1, \ldots, t_n)$ is free.

(ii) Variables in $\sim \mathbf{A}$, $\mathbf{A} \wedge \mathbf{B}$, $\mathbf{A} \vee \mathbf{B}$, $\mathbf{A} \Rightarrow \mathbf{B}$, and $\mathbf{A} \iff \mathbf{B}$ are free or bound according to whether they are free or bound in $\mathbf{A}$ or $\mathbf{B}$ separately.

(iii) The free and bound occurrences of a variable in a formula of the form $\exists x, \mathbf{A}$ or $\forall x, \mathbf{A}$ are the same as the free and bound occurrences of the variable itself, *except* that every occurrence of x is now considered bound.

Intuitively, we see that a variable is bound precisely when it is acted upon by a quantifier, $\forall$ or $\exists$. A *statement* is a formula that has no free variables.

2.4 The Semantics of First-Order Logic

2.4.1 Interpretations

If we are to assign a truth value to a formula A in first-order logic, then we must first give interpretations of the nonlogical symbols appearing in A. Here a nonlogical symbol is any symbol *other* than the universal ones listed in Subsection 2.1.1. The logical symbols are, of course, the ones inherent in the logic and not a part of any particular model.

Thus we must specify a domain or universe of objects and and must give meanings to each function symbol and relation symbol appearing in A. A structure or interpretation $\mathcal{M}$ for a given language L, then has the following components:

- a nonempty universe M of objects over which variables and terms range;
- for each k-ary function f in the language, an interpretation $f^{\mathcal{M}} : M^k \to M$;
- for each k-ary relation $\mathbf{P}$ in the language, an interpretation $P^{\mathcal{M}} \subseteq M^k$ containing all k-tuples for which we mean $\mathbf{P}$ to hold;
- in case the first-order language contains the equality symbol $=$, then the interpretation $=^{\mathcal{M}}$ must be the equality predicate on M.

2.4.2 Truth

If A is a sentence and $\mathcal{M}$ is a structure, then we write $\mathcal{M} \models A$ to mean that A is true in the structure $\mathcal{M}$. In this circumstance, we say that $\mathcal{M}$ is a *model* for A. We can also say that A is *satisfied* by $\mathcal{M}$.

If A is a formula (in which free variables occur), rather than a sentence (in which only bound variables occur), then we say that A is *valid* in $\mathcal{M}$ if $\mathcal{M} \models A[\sigma]$ for all object assignments σ of values in M to the variables in A. A formula is *valid* (without qualification) precisely when it is valid in all structures. Finally, for a formula A and set of formulas Γ, we write $\Gamma \models A$ (Γ tautologically implies A) to mean that A is valid in every structure in which Γ is valid.

2.4.3 First-Order Theories

Definition 2.1 A *first-order theory* is a set T of sentences that is closed under logical implication: If $T \models A$, then $A \in T$. An *axiomatization* of T is a set Γ of sentences such that T is precisely the set of sentences logically implied by Γ.

2.4.4 A Proof System for First-Order Logic

Following [BUS], we specify a proof system $\mathcal{F}_{FO}$ for first-order logic. It will consist of the ten logical axioms for the propositional calculus:

(i) $p \Rightarrow (q \Rightarrow p)$;

(ii) $(p \Rightarrow q) \Rightarrow [(p \Rightarrow\sim q) \Rightarrow\sim p]$;

(iii) $(p \Rightarrow q) \Rightarrow [(p \Rightarrow (q \Rightarrow r)) \Rightarrow (p \Rightarrow r)$;

(iv) $(\sim\sim p) \Rightarrow p$;

(v) $p \Rightarrow (p \vee q)$;

(vi) $(p \wedge q) \Rightarrow p$;

(vii) $q \Rightarrow (p \vee q)$;

(viii) $(p \wedge q) \Rightarrow q$;

(ix) $(p \Rightarrow r) \Rightarrow [(q \Rightarrow r) \Rightarrow ((p \vee q) \Rightarrow r)]$;

(x) $p \Rightarrow [q \Rightarrow (p \wedge q)]$;

together with two axiom schemes for the quantifiers

(xi) $A(t) \Rightarrow \exists x, A(x)$;

(xii) $\forall x, A(x) \Rightarrow A(t)$;

and two quantifier rules of inference

(xiii) $[C \Rightarrow A(x)] \Rightarrow [C \Rightarrow \forall x, A(x)]$;

(xiv) $[A(x) \Rightarrow C] \Rightarrow [\exists x, A(x) \Rightarrow C]$.

In both of the latter inferences, x may not appear as a free variable in C. If the first-order language being considered contains the $=$ relation, then these equality axioms must be included:

(xv) $\forall x, x = x$;

(xvi) $\forall \mathbf{x}, \forall \mathbf{y}, [x_1 = y_1 \wedge \cdots \wedge x_k = y_k] \Rightarrow [f(\mathbf{x}) = f(\mathbf{y})]$
(where $\mathbf{x} = (x_1, \ldots, x_k)$ and $\mathbf{y} = (y_1, \ldots, y_k)$);

(xvii) $\forall \mathbf{x}, \forall \mathbf{y}, [(x_1 = y_1 \wedge \cdots \wedge x_k = y_k) \wedge P(\mathbf{x})] \Rightarrow P(\mathbf{y})$
(where $\mathbf{x} = (x_1, \ldots, x_k)$ and $\mathbf{y} = (y_1, \ldots, y_k)$).

2.4.5 Two Fundamental Theorems

Theorem 2.1 (the soundness theorem)

(1) If $\mathcal{F}_{FO} \vdash A$, then $\models A$.

(2) Let Γ be a collection of sentences. If there is an $\mathcal{F}_{FO}$-proof of A using sentences from Γ as additional axioms, then $\Gamma \models A$.

The next theorem is a sort of converse to the first. It is, naturally, a version of Gödel's completeness theorem.

Theorem 2.2 (the completeness theorem)

(1) If $\models A$, then $\mathcal{F}_{FO} \vdash A$.

(2) Let Γ be a collection of formulas. If $\Gamma \models A$, then there is an $\mathcal{F}_{FO}$-proof of A using sentences from Γ as additional axioms.

Chapter 3

Axiomatics and Formalism in Mathematics

If every mathematician occasionally, perhaps only for an instant, feels an urge to move closer to reality, it is not because he believes that mathematics is lacking in meaning. He does not believe that mathematics consists in drawing brilliant conclusions from arbitrary axioms, of juggling concepts devoid of pragmatic content, of playing a meaningless game.

—Errett Bishop

The pursuit of knowledge is, I think, mainly actuated by love of power.

—Bertrand Russell

Logic is founded upon suppositions which do not correspond to anything in the actual world—for example, the supposition of the equality of things, and that of the identity of the same thing at different times.

—F.W. Nietzsche

The totality of thought is a picture of the world.

—Ludwig Wittgenstein

Logic is the art of going wrong with confidence.

—Joseph Wood Krutch

Logic is the art of convincing us of some truth.

—Jean de La Bruyère

Brouwer, who has done more for constructive mathematics than anyone else, thought it necessary to introduce a revolutionary, semimystical theory of the continuum.

—Errett Bishop

'Can you do addition?' the White Queen asked.
'What's one and one and one and one and one and one and one and one and one and one?'
'I don't know,' said Alice. 'I lost count.'

—Lewis Carroll, *Through the Looking Glass*

3.1 Basic Elements

3.1.1 Undefinable Terms

New terminology is defined in terms of old terminology. Any tract of logic contains only finitely many words. Therefore there must be a beginning, and that beginning cannot depend on anything that came previously. There must therefore be elementary terms that remain undefined. These terms are called "undefinable." We strive to keep the number of these terms to a minimum. Since the terms are formally undefinable, we must describe them in a manner that is heuristically self-evident.

In modern mathematics, the primary undefinable terms are "set" and "element of." A *set* is declared to be a collection of objects. We say that x is an element of the set S, and we write $x \in S$, if x is one of the objects that compose S. In case y is not an element of S, then we write $y \notin S$.

3.1.2 Description of Sets

If the elements of a set S are known explicitly and exhaustively, then we write them out in this way: $S = \{s_1, s_2, \ldots, s_k\}$. This set has k elements $s_1, s_2, \ldots, s_k$. More common in mathematics is that a set is defined by a condition, using "set-builder notation." As an example,

$$T = \{m \in \mathbb{Z} : m^2 - 3m + 5 > 0\}.$$

The colon in this expression is read "such that." We see that the set T is the collection of integers m that satisfy the inequality $m^2 - 3m + 5 > 0$.

Example 3.1

Let $S = \{p \in \mathbb{Q} : 4 < q \leq 9\}$. Then $6 \in S$, $9 \in S$, and $12 \notin S$.

□

3.1.3 Definitions

A *definition* gives the meaning of a term, or a piece of notation, or a concept. A definition can only use ideas that occurred earlier in the exposition. These earlier ideas could be previous definitions or they could be undefinables. (In more advanced mathematical exposition, a definition could depend on a theorem.) Given our discussion of "set" and "element of" in the previous subsection, we can now (for instance) present this definition:

Definition: Let A and B be sets. We say that A is a *subset* of B, and we write $A \subset B$, if $x \in A$ implies $x \in B$.

Observe that this definition explains the term "subset," and that it relies only on the concepts of "set" and "element of," and on universal terms of our logical system. All these have been treated beforehand. Another example is

> Let A and B be sets. We define the *intersection* of A and B, denoted $A \cap B$, to be
>
> $$A \cap B = \{x : x \in A \text{ and } x \in B\}.$$

Definitions take many forms. The simplest type of definition has the form

> **Definition**: A gaggle of ixnays with left-handed polyglot is called an *amscray*.

In mathematics and computer science, other more subtle types of definitions are sometimes useful. An *inductive definition* is one that defines a sequence of objects inductively. For example:

> **Definition**: Let the first and second Fibonacci numbers be $a_0 = 1$ and $a_1 = 1$. Inductively define $a_{j+2} = a_j + a_{j+1}$ for $j \geq 0$.

Inductive definitions of sets are often presented (informally) by giving some rules for generating elements of the set. Alternatively, one can specify that the set is the smallest set closed under the given rules. As a basic example, we can define the terms of a language to be the smallest set of expressions containing the variables and constants and closed under the "term formation rule":

> If $t_1, \ldots, t_n$ are terms and if f is an n-ary function symbol of the language, then the expression $f(t_1, \ldots, t_n)$ is a term.

Inductive definitions are useful when presenting the syntax of a formal language. There are also connections between inductive definitions and iteration of monotone operators.

In mathematics, it is often the case that a new theorem will make a definition possible. For example, the Fundamental Theorem of Algebra says that every nonconstant polynomial with complex coefficients has a complex root. With this knowledge, one can use the Euclidean algorithm to show that a polynomial of degree k has a factorization into k (not necessarily distinct) factors. Then the following definition is possible:

> **Definition**: Let p be a polynomial of degree $k > 1$ with complex coefficients. Then p may be written as
>
> $$p(z) = c \cdot (z - a_1)^{\ell_1} \cdot (z - a_2)^{\ell_2} \cdots (z - a_m)^{\ell_m},$$

where $a_1, \ldots, a_m$ are the (distinct) roots of p with multiplicities $\ell_1, \ldots, \ell_m$, respectively, and $\ell_1 + \cdots + \ell_m = k$.

We have mentioned definitions that are used to defined terminology, and we have mentioned definitions that are used to define concepts. There are also definitions that are used to define notation. An example is the following.

Definition (Einstein Summation Notation): In our displayed formulas, the expression

$$a_j b^j$$

with the index j repeated—once as a subscript and once as a superscript—will denote the sum

$$\sum_{j=1}^{4} a_j b^j .$$

It is important that a definition only use terminology, concepts, and results that have occurred *before* the definition. A definition should not include—on the fly—the subdefinition of a needed term. It should also not include the quick and dirty enunciation of some preliminary result. All of those items should be laid out *before* the definition is formulated.

3.1.4 Axioms

The idea of axiom is to the concept of theorem as an undefinable is to a definition. That is to say, an axiom is an assertion about relationships among the ideas introduced, but it is one that we cannot prove. For example, in a rigorous development of Euclidean geometry we treat "point" and "line" and "betweenness" and "incidence" as undefinables. Then we enunciate these axioms:

P1 Through any pair of distinct points there passes a line.

P2 For each segment $\overline{AB}$ and each segment $\overline{CD}$, there is a unique point E (on the line determined by A and B) such that B is between A and E and the segment CD is congruent to BE.

P3 For each point C and each point A distinct from C, there exists a circle with center C and radius CA.

P4 All right angles are congruent.

P5 For each line ℓ and each point $\mathbf{P}$ that does not lie on ℓ there is a unique line m through $\mathbf{P}$ such that m is parallel to ℓ.

There is no sense, nor any meaning, in trying to prove these axioms. They are given, and all other assertions of Euclidean geometry are derived from these. As with an undefinable, an axiom should be heuristically appealing and, as much as possible, self-evident.

There are two types of axioms: logical axioms and nonlogical axioms. A *logical axiom* is one that is an artifact of the logic we use, and is not particular to the specific language (such as traditional, rectilinear Euclidean geometry) being considered. A nonlogical axiom is one that is an artifact of the language. Logical axioms are syntactical, and nonlogical axioms are semantic. Euclid's axioms are plainly syntactical, as they apply to many different geometries. In other words, these axioms are about *logical combinations of the undefinables.* They are *not* model-dependent.

An axiom should only use terms that are defined *previous* to the statement of the axiom. Usually an axiom is not dependent (either logically or for understanding) on a theorem or proposition, although the rules of logic do not rule out the possibility.

3.1.5 Lemmas, Propositions, Theorems, and Corollaries

Lemmas, propositions, theorems, and corollaries are all statements of fact that we derive, using the rules of logic, from the axioms and from previous definitions, axioms, lemmas, propositions, theorems, and corollaries. The four different names indicate relative importance and sometimes logical dependence. Specifically:

- A lemma is a result that has no intrinsic importance but is needed in order to establish some more important fact. The lemma is isolated simply to clarify the logic and to help the exposition.

- A proposition is more important than a lemma. It will have some intrinsic interest but will not be a fundamental result.

- A theorem is a major assertion, a milestone in the subject being explicated. The proof of the theorem may depend on several preceding lemmas and propositions.

- A corollary is a consequence of some theorem or proposition—one that may be derived rather immediately as a by-product of ideas already presented.

The following statements can be proved as exercises in formal logic:

- Any axiom can be proved as a theorem.

- Any theorem can be added to the list of axioms.

The statement of a theorem should be as clear and as *concise* as possible. All the necessary definitions and prior results (lemmas and so forth) should be recorded and explained *before* the theorem is formulated.

Brevity is important in the statement of a theorem. If your theorem has 23 hypotheses and 17 conclusions, then you should either (i) introduce terminology that will amalgamate groups of the hypotheses into umbrella concepts or (ii) break the theorem up into component results.

3.1.6 Rules of Logic

The primary rule of logic is *modus ponendo ponens* (or *modus ponens* for short).[1] It says

If A and $[A \Rightarrow B]$, then B.

Some treatments also formulate *modus tollens*, which is in fact logically equivalent to *modus ponens*. It says

If $\sim B$ and $A \Rightarrow B$, then $\sim A$.

Modus tollens actualizes the fact that $\sim B \Rightarrow \sim A$ is logically equivalent to $A \Rightarrow B$. The first of these implications is called the *contrapositive* of the second.

The Axiom of Substitution (Axiom Schema of Replacement—Subsection 4.2.7) is also an important rule of logical reasoning. We will say more about it later.

In a formal treatment of proof theory (see [BUS, p. 5 ff.]), we specify—in addition to *modus ponens*—a system of logical axioms that allow the inference of "self-evident" tautologies from no hypotheses. See Subsection 2.4.4 for the enunciation of such a system.

Let us refer to that system of ten logical axioms as $\mathcal{F}$, in honor of Gottlob Frege (1848–1925). It is a remarkable fact that $\mathcal{F}$ is complete in the sense that any tautological statement of the propositional calculus can be proved using $\mathcal{F}$. See also Section 1.4 on the Gödel completeness theorem.

It is an open question whether proof systems that allow substitution are actually more powerful than the proof system $\mathcal{F}$ + (*modus ponens*). Certainly substitution allows for more elegant and more readable proofs. We refer the reader to [BUS] for further details.

The question of finding the best algorithmic methods for propositional proofs is currently an active area of research. It is important

[1] In computer science contexts, *modus ponens* is sometimes called "implies-elimination" or "arrow-elimination." See also the concept of cut elimination, which is discussed in Section 10.2.

for artificial intelligence, for automated theorem proving, and for logical programming.

3.1.7 Proofs

A *proof* is a sequence of statements that lead from the definitions and the axioms to some statement that we wish to establish. The only allowable rule for moving from the nth statement to the $(n + 1)$st statement is to apply *modus ponens* (or, equivalently, *modus tollens*) or the Axiom Schema of Replacement. Along the way, we may set lemmas, propositions, corollaries, and theorems as milestones. But those are only labels. The *proof* is nothing other than a sequence of logical steps that lead to the desired result. Chapter 8 discusses different methods of proof.

3.2 Models

3.2.1 Definition of Model

A *model* for a collection S of statements is a set M together with an interpretation of some constant and relation symbols. This "interpretation" consists of a map $c_\alpha \mapsto \overline{c}_\alpha$ from the constant elements of S to elements of M and a second map $R_\beta \mapsto \overline{R}_\beta$ from the relation elements of S to elements of M such that all the statements of S are true in M.

3.2.2 Examples of Models

Let us consider classical rectilinear geometry. The standard axiom system for geometry is given by the first four axioms enunciated in Subsection 3.1.4. This set of four axioms is a set S of syntactic statements. A *model* (indeed, the standard model) for S is the usual Euclidean geometry consisting of points and lines in the plane with their usual betweenness and incidence relations. If we replace the plane by the sphere and the points by points on the sphere and the lines by great circles on the sphere, then we obtain a new model for geometry (see [GRE] or [KRA] for details). (Another approach to obtaining a noneuclidean geometry is to use the hyperbolic geometry of the disc—see [GRE].)

Now let us add a fifth axiom. If we utilize the fifth axiom **P5** as enunciated in Subsection 3.1.4, then the model with points and lines in the plane will indeed satisfy this five-axiom geometry. If we instead utilize the *negation* of the fifth axiom **P5**, then the model with great circles in the sphere will satisfy this new five-axiom geometry. The first of these geometries is called *Euclidean*. The second is called *noneuclidean*. These are different geometries, because they differ in the truth value of the fifth axiom (the celebrated "parallel postulate").

3.2.3 Finite Model Theory

"Finite model theory" is a part of modal logic. One of the fundamental ideas is the following definition.

Definition 3.1 . A modal logic **L** (Subsection 1.2.3) has the *finite model property* if, for each formula X that is not valid in **L**, there is a finite model in which X is false.

The reader will see that the spirit of this definition is inspired by the classical compactness theorems (see Subsection 1.4.6). Now, a basic result about finite models is the following proposition.

Proposition 3.1

If a modal logic **L** *has a proof procedure and has the finite model property, then* **L** *is decidable.*

Here a system S is *decidable* if, for any statement **A** in S, either **A** or $\sim$ **A** can be proved in S.

The typical device for proving that a particular logic has the finite model property is a filtration. The basic idea of a filtration is to identify a sequence of "worlds" (i.e., pairs of truth-functional assignments and sets of propositions) that sort out the formulas on which we wish to focus. The full idea is explained in [GHR], which contains a more complete treatment of finite models.

3.2.4 Minimality of Models

Let S be a set of statements. A model M for S is said to be *minimal* if it has least possible cardinality. The Löwenheim–Skolem theorem (in first-order logic) says that every model for a collection T of constants and relations has an elementary submodel whose cardinality does not exceed the cardinality of T or (in case T is finite) is countable. (In another context, Section 1.4.4, we have called such a model "coordinated.") Clearly, the model provided by Löwenheim–Skolem is coordinated, and it is minimal.

A complementary result is Tarski's cardinality theorem: Let $\mathbf{m}$ be an infinite cardinal. Let T be an $\mathbf{m}$-theory (i.e., a theory whose language has a set of nonlogical symbols having cardinality less than or equal to $\mathbf{m}$) having an infinite model. Then T has a model of cardinality $\mathbf{m}$.

3.2.5 Universal Algebra

A useful technical device in model theory is *universal algebra.* To put the matter simplistically, universal algebra abstracts the notions of a group, ring, or field to an algebraic system with certain n-ary (or even infinitary) operations. It turns out that any logical theory can be regarded as an instance of universal algebra: the elements of the algebra are the theorems (i.e., the provable statements in the theory), and the operations are the usual rules of deduction. The compactness theorem and Gödel's completeness theorem can be elegantly understood in the context of universal algebra. A nice reference is [BRI].

3.3 Consistency

3.3.1 Definition of Consistency

A collection of statements S is *consistent* if the statement $A \wedge \sim A$ cannot be derived from it for any A. The axioms of set theory, the axioms of the real numbers, the axioms of group theory, and the axioms of a field—indeed all the standard axiom systems of mathematics—are known to be consistent. A standard method for proving a collection S to be consistent is to produce a model.

It is a fairly recent result of modern mathematics (Gödel, 1939) that the axioms of set theory with the continuum hypothesis adjoined are consistent.

3.4 Gödel's Incompleteness Theorem

3.4.1 Introductory Remarks

The celebrated incompleteness theorem of Gödel asserts that any logical system that is sufficiently complex to contain arithmetic will be incomplete. This "incompleteness" means that there are statements A that can be formulated in the system so that neither A nor $\sim A$ can be proved in the system.

The statement is shocking in its formulation because it says that even elementary systems are not completely resolvable. In particular, even elementary computer languages will contain statements that can neither be established nor counter-proved in finitely many steps.

3.4.2 Gödel's Theorem and Arithmetic

The reason for requiring the system to be sufficiently complex to contain arithmetic is that Gödel's proof relies in an essential way on an enumeration of all provable statements in the logical system. In the classical terminology, we assign a "Gödel number" to each provable statement in the system. Then we use a diagonalization argument, very much in the spirit of Georg Cantor's original diagonalization procedure (Subsection 5.8.2), to produce a nonprovable statement. A formal enunciation of Gödel's theorem will follow.

3.4.3 Formal Enunciation of Arithmetic

First, we give an explicit formulation of arithmetic, which will be referred to in the sequel as Z_1. (Here we use the standard notation ! to mean "unique.")

1. $\forall x, y \exists! z, x + y = z;$

2. $\forall x, y \exists! z, x \cdot y = z;$

3. $\forall x, x + 0 = x \ \wedge \ x \cdot 1 = x;$

4. $\forall x, y, (x + (y + 1) = (x + y) + 1;$

5. $\forall x, y, x + 1 = y + 1 \Rightarrow x = y;$

6. $\forall x, \sim (x + 1 = 0).$

3.4.4 Some Standard Terminology

The following standard terminology will be used in the formulation of Gödel's results:

consistent: A set S of statements is said to be *consistent* if the statement $A \wedge \sim A$ cannot be derived from it for any A. We will write Consis S. If S has a model, then it is consistent.

complete: A system is *complete* if every true statement in the system is provable in that system. (Of course it is obvious that every provable statement is true.)

primitive recursive: A function $f(n_1, \ldots, n_k)$ from $\mathbb{Z}$ to $\mathbb{Z}$ is called *primitive recursive* (p.r.) if it is constructed by means of the following rules.

1. $f \equiv c$ for some constant c is p.r.
2. $f(n_1, \ldots, n_k) = n_i$, some $1 \leq i \leq k$, is p.r.
3. $f(n) = n + 1$ is p.r.
4. If $f(n_1, \ldots, n_k)$ and $g_1, \ldots, g_k$ are p.r., then so is $f(g_1, \ldots, g_k)$.

5. If $f(0, n_2, \ldots, n_)$ is p.r., if $g(m, n_1, \ldots, n_k)$ is p.r., and if we have $f(n_1, n_2, \ldots, n_k) = g(f(n, n_2, \ldots, n_k), n, n_2, \ldots, n_k)$, then f is p.r.

general recursive: A function $f(x_1, \ldots, x_k)$ is *general recursive* (or just recursive) if there is a finite set of equations (in terms of f) such that, for any choice of the numerals $n_1, \ldots, n_k$, there is a unique m such that $f(n_1, \ldots, n_k) = m$ can be deduced.

recursively enumerable: A set S is *recursively enumerable* if S is either empty or is the image of a general recursive function.

3.4.5 Enunciation of the Incompleteness Theorem

Now we have the incompleteness theorem.

Theorem 3.1
Let Z_1 be the foregoing formulation of arithmetic. Then there is a statement A in Z_1 such that neither A nor $\sim A$ can be proved from the axioms of Z_1.

A version of the incompleteness theorem that is commonly invoked in computer science contexts is the following.

Theorem 3.2
Consis Z_1 cannot be proved in Z_1.

More generally, we have the following theorem.

Theorem 3.3
Let Σ be a formal system whose axioms are given by some recursive rule. If Σ is consistent, and if the primitive recursive functions can be embedded in Σ, then Consis Σ cannot be proved in Σ.

3.4.6 Church's Theorem

Church formulated the following result that puts the Gödel incompleteness theorem into a context.

Theorem 3.4 (Church)
If T is a consistent extension of the Peano theory of the natural numbers, then T is undecidable (see Section 3.5).

There are many variants and reformulations of Church's theorem. We now enunciate two of them.

Theorem 3.5
No decision procedure exists for arithmetic.

Theorem 3.6
In first-order logic, the theorems are recursively enumerable but the non-theorems are not.

3.4.7 *Additional Formulations of Incompleteness*

We take this opportunity to formulate some other versions of the incompleteness theorem that are commonly encountered. Note in passing that a formal system T is *ω-consistent* if there is no formula W with one free variable such that $W(n)$ is a theorem for every natural number n but there is some x such that $W(x)$ is not a theorem. The property of ω-consistency is stronger than the property of consistency.

Theorem 3.7 (first incompleteness theorem)
Let T be a formal theory containing arithmetic. Then there is a sentence φ that asserts its own unprovability and is such that:

(i) If T is consistent, then T does not prove φ.

(ii) If T is ω-consistent, then T does not prove $\sim \varphi$.

Theorem 3.8 (second incompleteness theorem)
Let T be a consistent formal theory containing arithmetic. Then T does not tautologically imply Consis T.

It is this second incompleteness theorem that enunciates the principle that no system (of a sufficient minimal level of complexity) can be established to be consistent without reference to an outside system. Thus we have the notion of *relative consistency.*

3.4.8 *Relative Consistency*

Gödel's incompleteness theorem establishes the importance of relative consistency. His theorem, in effect, says that the completeness of a system (of certain minimal complexity) cannot be established *within that system.* Thus we typically show that a system S is complete *relative to* some other system T.

As an example, Paul Cohen proved that adding the denial of the continuum hypothesis ($\sim CH$) to the other axioms of Zermelo–Fraenkel set theory (*ZF*) does not lead to a contradiction, *unless the axioms of ZF themselves are already contradictory.* Thus $\sim CH$ is consistent relative to the axioms of *ZF*.

3.5 Decidability and Undecidability

3.5.1 Introduction to Decidability

Briefly, a system *S* is *decidable* if, for any statement **A** in *S*, either **A** or $\sim$ **A** can be proved in *S*. Here "proved" means that there is a sequence of *finitely many* logical steps leading from the axioms to the statement being proved. The system is *undecidable* if there is a statement **A** for which no such proof of either **A** or $\sim$ **A** can be found.

The question of decidability is very closed related to the concept of recursion. For example, *Church's thesis* asserts that the collection of general recursive functions exhausts the class of "effectively computable" functions. This is a philosophical statement, not one that is amenable to proof. An effectively computable function is a concrete form of a decidable statement.

3.5.2 Recursive Equivalence; Degrees of Recursive Unsolvability

The set of recursive functions (see Chapter 6) may be stratified in the following way. Say that a function F is *recursive in the function* G if F may be built from G using the steps allowed in inductively forming a recursive function. Define a relation on functions by saying that $F \sim G$ if both F is recursive in G and G is recursive in F. This is in fact an equivalence relation (Section 5.4). The resulting equivalence classes are called *degrees of recursive unsolvability*, or simply *degrees.* Intuitively, two functions are in the same equivalence class if they are equally difficult (or equally complex) to calculate. For more on this matter, see [SCH, p. 169].

3.6 Independence

3.6.1 Introduction to Independence

Let S be a finite set of statements. We say that the statements in S are *independent* if no one of them is implied by the others. Now let **A** be a statement that is not in S. We say that **A** is independent of S if S does not imply **A**. Alternatively, **A** is independent of S if both **A** and $\sim$ **A** are consistent with S.

The standard (though not the only) method for confirming that **A** is independent of S is to construct a model M_1 in which S is true and **A** is true, and then to construct a second model M_2 in which S is true but **A** is false.

3.6.2 Examples of Independence

Three of the most standard, and celebrated, examples of independence are the following.

(1) Let S consist of the first four axioms **P1**, ..., **P4** of geometry, and let **A** be **P5**, the parallel postulate. See Subsection 3.1.4. Then the standard planar geometry of points and lines is a model in which S is true and **A** is true. But the hyperbolic geometry of the disc, or the geometry of great circles on the sphere, is a model in which S is true but **A** is false. Thus the parallel postulate is independent of the other axioms of geometry.

(2) Let S consist of the axioms of set theory as enunciated in Section 4.2. Let *CH* be the continuum hypothesis. Then Gödel [GOD] has produced a model for set theory (the "constructible sets") in which both S and *CH* are true. Paul Cohen [COH] has produced another model for set theory (using the technique of "forcing") in which S is true and *CH* is false. Thus the continuum hypothesis is independent of the other axioms of set theory.

(3) In the "constructible" model of set theory mentioned in (2), the Axiom of Choice (*AC*) is true. Paul Cohen used forcing to construct a model of set theory in which *AC* is false. Thus the Axiom of Choice is independent of the other axioms of set theory.

Chapter 4

The Axioms of Set Theory

There are two kinds of arguments, the true and the false. The young should be instructed in both—but the false first.

—Plato

Mathematics takes us still further from what is human, into the region of absolute necessity, to which not only the actual world, but every possible world, must conform.

—Bertrand Russell

Logic is logic. That's all I'll say.

—O.W. Holmes

Logic is neither a science nor an art, but a dodge.

—Benjamin Jowett

Error has its logic as well as truth.

—George Plechanoff

The study of mathematics, like the Nile, begins in minuteness, but ends in magnificence.

—C.C. Colton

What can be said can be said clearly, and what we cannot talk about we must consign to silence.

—Ludwig Wittgenstein

Logic is nothing more than a knowledge of words.

—Charles Lamb

Structures are the weapons of the mathematician.

—Nicholas Bourbaki

The human mind has to first construct forms, independently, before we can find them in things.

—*Albert Einstein*

4.1 Introduction

Of course there are many versions of set theory, and the inclusion or exclusion of certain axioms is the subject of intensive study. Here we present one of the most standard versions of set theory, adhering closely to the version first set down by Zermelo and Fraenkel.

As you read these axioms, bear in mind that part of the philosophy behind them is to limit the ways in which new sets can be formed from old ones. The formulators of these axioms were thinking of ways to circumvent Russell's Paradox (Subsection 5.9.1) and other fundamental conundra of set theory.

4.2 Axioms and Discussion

We treat each axiom in its own subsection, and provide a brief discussion of meaning and context in each case.

4.2.1 Axiom of Extensionality

Axiom I:

$$[\forall x, (x \in A \iff x \in B)] \Rightarrow (A = B).$$

The Axiom of Extensionality says that two sets are equal if and only if they have the same elements.

4.2.2 Sum Axiom

Axiom II: Given a collection $\mathcal{S}$ of sets,

$$\exists C, \forall x, \Big(x \in C \iff \exists B, [x \in B \land B \in \mathcal{S}]\Big).$$

The Sum Axiom says that, given any collection of sets, there exists a set that is their union. A variant of this axiom is known as the Union Axiom ([SUP2, p. 34]).

4.2.3 Power Set Axiom

Axiom III: Given a set A,

$$\exists B, \forall C, \big(C \in B \iff C \subset A\big).$$

The Power Set Axiom says that, given any set A, there is a set that is the power set of A. An alternative approach to these ideas is to use the Pairing Axiom ([SUP2, p. 31]).

4.2.4 Axiom of Regularity

Axiom IV:

$$A \neq \emptyset \Rightarrow \exists x, \left[x \in A \land \forall y, (y \in x \Rightarrow y \notin A)\right].$$

The Axiom of Regularity rules out sets that are elements of themselves. Combined with the other axioms, it also rules out infinite descending "element of" relations ([SUP2, p. 53]). Versions of this postulate are also called the Foundation Axiom ([SUP2, p. 59]).

4.2.5 Axiom for Cardinals

Axiom V: If A is a set, then let $\mathcal{K}(A)$ denote the collection of all sets that are set-theoretically isomorphic to A (i.e., sets that have the same cardinality as A—see Subsection 5.7.5). The set $\mathcal{K}(A)$ is sometimes called the *cardinality* of A, or the *cardinal number* corresponding to A. Let $\equiv$ denote set-theoretic isomorphism. Then

$$\mathcal{K}(A) = \mathcal{K}(B) \iff A \equiv B.$$

The Axiom for Cardinals specifies that two sets have the same cardinality if and only if they belong to the same cardinal number (i.e., same equivalence class under set-theoretic isomorphism). In other words, the axiom allows us to create the set that is the cardinal number.

4.2.6 Axiom of Infinity

Axiom VI:

$$\exists A, \left(\emptyset \in A \land \forall B, [B \in A \Rightarrow B \cup \{B\} \in A]\right).$$

The Axiom of Infinity specifies the existence of an infinite set. To understand this point, notice that it is impossible for a finite set to be set-theoretically isomorphic to a proper subset of itself. The Axiom of Infinity says that there exists some set that *is* set-theoretically isomorphic to a proper subset of itself. That set must therefore be infinite.

4.2.7 Axiom Schema of Replacement

Axiom VII: Let $P(x, y)$ be a property of x and y. If

$$\forall x, \forall y, \forall z, \left[\left(x \in A \land P(x,y) \land P(x,z)\right) \Rightarrow (y = z)\right],$$

then

$$\exists B \forall y, \left[y \in B \iff \exists x, (x \in A \land P(x,y))\right].$$

The Axiom Schema of Replacement specifies that, for a suitable property $\mathbf{P}$ (satisfying the first condition), a new set may be defined using that property $\mathbf{P}$. Although this axiom is perhaps the most technical of the eight, it is the one that we most frequently use explicitly; it is the standard device for forming new sets from old ones. This axiom is one of the basic logical rules. An alternative view of this axiom schema is the Axiom of Separation ([SUP2, pp. 6 ff.]).

4.2.8 Axiom of Choice

Axiom VIII: Let A be any set. Let $\mathcal{P}(A)$ be the power set (i.e., the set of all subsets of A). Then there is a function $f : \mathcal{P}(A) \to A$ such that, for any nonempty subset B of A, $f(B) \in B$.

The Axiom of Choice gives us the power to choose one element from each of the subsets of a given set A. In case A is countable, this matter may sometimes be handled in an *ad hoc* fashion (but see the Bertrand Russell quotation at the beginning of Chapter 9). For uncountable A, the Axiom of Choice is a powerful tool and can also lead to surprising results (see Sections 9.3 and 9.4).

4.3 Concluding Remarks

In Chapter 5, we discuss the elements of set theory. In a more formal treatment we would specify how each part of the discussion depends on the eight axioms listed above. We will not engage in such a rigorous exercise in the present book (instead see [SUP2]). In Chapter 9, we give a more thorough treatment of the Axiom of Choice. A detailed and rigorous treatment of the other axioms is best left for a formal course in set theory and logic.

Chapter 5

Elementary Set Theory

I recollect an acquaintance saying to me that "the Oriel Common Room stank of Logic."

—John Henry, Cardinal Newman

Since argument is not recognized as a means of arriving at truth, adherents of rival dogmas have no method except war by means of which to reach a decision. And war, in our scientific age, means, sooner or later, universal death.

—Bertrand Russell

If logic is the hygiene of the mathematician, it is not his source of food.

—André Weil

Logic, like whiskey, loses its beneficial effect when taken in too large quantities.

—Lord Dunsany

If we take in our hand any volume; of divinity or school metaphysics, for instance; let us ask: Does it contain any abstract reasoning concerning quantity or number? No. Does it contain any experimental reasoning, concerning matter of fact and existence? No. Commit it then to the flames: for it can contain nothing but sophistry and illusion.

—David Hume

Mathematics contains much that will neither hurt one if one does not know it nor help one if one does know it.

—H.L. Mencken

And for mathematical sciences, he that doubts their certainty hath need of a dose of hellebore.

—Joseph Glanvill

That which mirrors itself in language, language cannot represent.

—Ludwig Wittgenstein

Logic is the anatomy of thought.

—John Locke

5.1 Set Notation

5.1.1 Elements of a Set

As indicated in our section on axiomatics (Chapter 3, especially Section 3.1.1), "set" is one of our undefined terms. We simply say that a set is a collection of objects. Examples are:

- the set of all footballs;
- the set of all monogamous ocelots;
- the set of all real numbers;
- the set of all ordered pairs of integers with first entry positive.

We generally use an uppercase roman letter to denote a set. If A is a set and a is one of the "objects" in the collection denoted by A, then we write $a \in A$. This expression is read "a is an element of A." In case a is not one of the objects in A, or not an element of A, then we write $a \notin A$.

5.1.2 Set-Builder Notation

In the mathematical sciences, the most common way to specify a set is with "set-builder notation" (see the Axiom Schema of Replacement in Subsection 4.2.7). Set-builder notation is predicated on the notion that a new set is defined relative to an old set. For example, we are familiar with the set of integers, and we commonly denote it by $\mathbb{Z}$. Thus we might define

$$S = \{m : m \in \mathbb{Z} \text{ and } m > 3\}.$$

The description of S is read

> The set of all m such that m is an element of the integers and m is greater than 3.

In particular, the colon is read as "such that." It is sometimes convenient to use a vertical bar $|$ instead of the colon.

Another example of set-builder notation is

$$A = \{x \in \mathbb{R} : x^2 + 3x + 1 = 0\}.$$

Thus A is the set of real numbers that satisfy a certain polynomial equation.

5.1.3 The Empty Set

One distinguished set that will come up in our discussions is the *empty set*. The empty set is the set with no elements. It is denoted by $\emptyset$. Thus it is never true that $x \in \emptyset$ for any x whatsoever.

5.1.4 Universal Sets and Complements

If it is understood from context that all sets being discussed are subsets of some universal set X, and if $A \subset X$, then we use the notation ${}^cA$ to denote the *complement* of A; the complement of A is the set of elements in the universe and not belonging to A. More rigorously, ${}^cA \equiv \{x \in X : x \notin A\}$.

Example 5.1

Suppose that we are discussing sets of real numbers, so $X = \mathbb{R}$. Let $A = \{x \in \mathbb{R} : -2 \leq x < 4\}$, $B = \{1, 2, 3, 5\}$, and $C = \{x \in \mathbb{Q} : 1 < x < \pi\}$. Then

$$\begin{aligned}
{}^cA &= \{x \in \mathbb{R} : x < -2 \text{ or } x \geq 4\} \\
{}^cB &= \{x \in \mathbb{R} : x < 1 \text{ or } 1 < x < 2 \text{ or } 2 < x < 3 \\
&\qquad \text{or } 3 < x < 5 \text{ or } 5 < x\} \\
{}^cC &= \{x \in \mathbb{R} : x \leq 1 \text{ or } [1 < x < \pi \text{ and } x \text{ is irrational}] \\
&\qquad \text{or } x \geq \pi\}.
\end{aligned}$$

□

5.1.5 Set-Theoretic Difference

If A and B are sets, then their *set-theoretic difference* is $A \setminus B \equiv \{x : x \in A \text{ and } x \notin B\}$. Later, after we define the notion of intersection of sets, we will have another way to understand the set-theoretic difference. We now illustrate with an example.

Example 5.2

Let $X = \{x \in \mathbb{R} : 2 \leq x < 4\}$, $Y = \{y \in \mathbb{R} : 3 < y \leq 5\}$, and $Z = \{z \in \mathbb{R} : 0 \leq z \leq 6\}$. Then

$$\begin{aligned}
X \setminus Y &= \{x \in \mathbb{R} : 2 \leq x \leq 3\}; \\
Y \setminus X &= \{y \in \mathbb{R} : 4 \leq y \leq 5\}; \\
X \setminus Z &= \emptyset; \\
Z \setminus X &= \{z \in \mathbb{R} : 0 \leq z < 2 \text{ or } 4 \leq z \leq 6\}; \\
Y \setminus Z &= \emptyset; \\
Z \setminus Y &= \{z \in \mathbb{R} : 0 \leq z \leq 3 \text{ or } 5 < z \leq 6\}.
\end{aligned}$$

□

5.1.6 Ordered Pairs; the Product of Two Sets

If S and T are sets, then the set of all *ordered pairs* of elements of S and T is denoted by $S \times T$. This is called the *set-theoretic product* of S and T. More rigorously, the set of such ordered pairs is the set $\{\{s, \{s,t\}\} : s \in S, t \in T\}$. We usually find it more convenient to denote an ordered pair by (s, t).

Example 5.3

Let $S = \{a, b, c\}$ and $T = \{1, 2, 3, 4\}$. Then

$$S \times T = \{(a,1),(a,2),(a,3),(a,4),(b,1),(b,2),(b,3),(b,4),\\ (c,1),(c,2),(c,3),(c,4)\}.$$

By contrast,

$$T \times S = \{(1,a),(1,b),(1,c),(2,a),(2,b),(2,c),\\ (3,a),(3,b),(3,c),(4,a),(4,b),(4,c)\}.$$

□

5.2 Sets, Subsets, and Elements

5.2.1 The Elements of a Set

We have already specified that $\in$ denotes the concept of "element of." Using the sets defined in Subsection 5.1.2, we see the "element of" relations $6 \in S$ and $[-3+\sqrt{5}]/2 \in A$. We now define a relationship between two sets that is called "containment" or "subset of."

> Let A and B be sets. We say that A is contained in B, or A is a subset of B, and we write $A \subset B$, if
>
> $$\forall x, x \in A \Rightarrow x \in B.$$

In other words, $A \subset B$ if every element of A is also an element of B. An alternative terminology is to say that "A is a subset of B." Every set A is a subset of itself, so $A \subset A$ is always true. In case $A \subset B$ and $A \neq B$, then we say that A is a *proper subset* of B. For emphasis, this is sometimes written $A \underset{\neq}{\subset} B$. Some books and tracts use $A \subseteq B$ for arbitrary set containment and $A \subset B$ for proper containment. We will use $\subset$ and $\subseteq$ interchangeably; this is the most prevalent custom in mathematics.

If C is *not* a subset of D, then we write $C \not\subset D$. This means that there is an element of C that is not an element of D.

Example 5.4

Let

$$A = \{2,4,6,8\},\ B = \{1,2,3,4,5,6,7,8\},\ C = \{1,3,5,7\}$$

$$D = \{1,2,3,4\},\ E = \{7,8,9,10\},\ F = \{1,2,3\}.$$

Then, by inspection, we find that

$$A \subset B,\ C \subset B,\ D \subset B$$

$$F \subset D,\ F \subset B.$$

There are no other containment relations. We may write

$$E \not\subset B,\ C \not\subset A,$$

and so forth. There are too many noncontainment relations to list them all.

□

Example 5.5

Let

$$X = \{x \in \mathbb{R} : -3 < x < 8\},\ Y = \{y \in \mathbb{Z} : x^2 - 3x + 2 = 0\}$$

$$Z = \{z \in \mathbb{Q} : 3/2 \leq z < 7\}.$$

Then

$$Y \subset X \quad \text{and} \quad Z \subset X.$$

There are no other containment relations.

□

5.2.2 Venn Diagrams

We sometimes use a Venn diagram to illustrate set containment. Figure 5.1 is a Venn diagram that depicts the containment relations among X, Y, Z that we discussed in the last example. We see that each set is represented by a planar region, the subset relation is illustrated by containment of regions, and intersection is illustrated by overlap of regions. (In Venn's original conception, each set was represented by a circle or disc. The modern custom is to use arbitrary planar regions to represent sets in a Venn diagram.) We will use Venn diagrams below to illustrate more sophisticated concepts such as intersection and union.

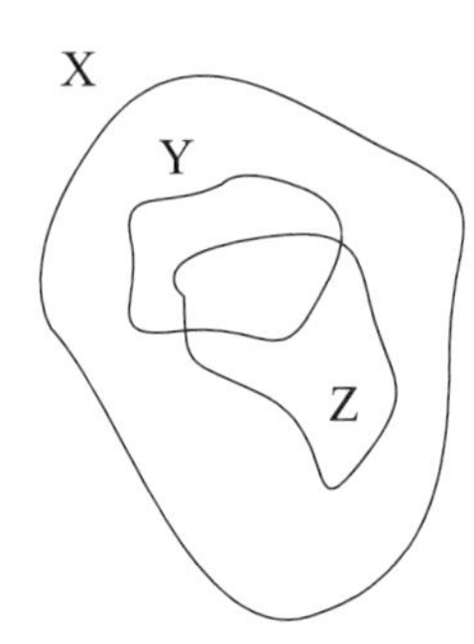

Figure 5.1

5.3 Binary Operations on Sets

5.3.1 Intersection and Union

Let X and Y be sets. We let $X \cup Y$ denote the set consisting of all elements of X and all elements of Y. The set $X \cup Y$ is called the *union* of the sets X and Y. More rigorously,

$$a \in X \cup Y \iff [a \in X \vee z \in Y].$$

Observe that we use $\vee$ here in the strictly rigorous sense of Sections 1.1 and 1.2. See also the mention of the Union Axiom in Subsection 4.2.2.

We also define the *intersection* of the sets X and Y to be the set of elements of both X and Y. We denote the intersection by $X \cap Y$. More rigorously,

$$a \in X \cap Y \iff [a \in X \wedge a \in Y].$$

Here we use $\wedge$ in the strictly rigorous sense of Sections 1.1 and 1.2.

Example 5.6

Let $C = \{x \in \mathbb{R} : 3 < x < 6\}$ and $D = \{y \in \mathbb{R} : 5 \le y \le 9\}$. Then

$$C \cup D = \{z \in \mathbb{R} : 3 < z \le 9\}$$

and

$$C \cap D = \{z \in \mathbb{R} : 5 \le z < 6\}.$$

□

Example 5.7

Let $X = \{m \in \mathbb{Z} : 1 < m^2 < 36\}$, $Y = \{x \in \mathbb{R} : x - 1 < 3\}$, and $Z = \{w \in \mathbb{Q} : -2 < w < 8\}$. Then

$$X \cap Y = \{m \in \mathbb{Z} : -6 < m < -1 \text{ or } 1 < m < 4\},$$

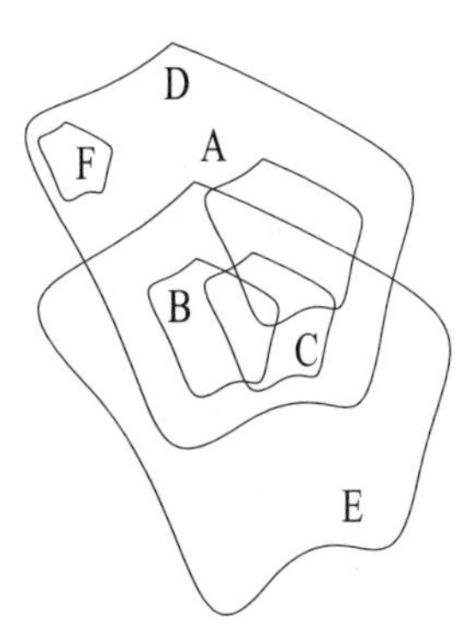

Figure 5.2

$$Y \cap Z = \{w \in \mathbb{Q} : -2 < w < 4\},$$

and

$$X \cap Z = \{m \in \mathbb{Z} : 1 < m < 6\}.$$

Also

$$X \cup Y = \{x \in \mathbb{R} : x < 4\} \cup \{4, 5\},$$

$$X \cup Z = \{-5, -4, -3, -2\} \cup \{w \in \mathbb{Q} : -2 < w < 8\},$$

and

$$Y \cup Z = \{x \in \mathbb{R} : x < 4\} \cup \{4, 5, 6, 7\}.$$

□

We conclude this subsection by using a Venn diagram to illustrate intersection, union, and containment.

Example 5.8

Let the universe X be the real numbers. Let

$$A = \{x \in \mathbb{R} : -2 < x < 5\},$$

$$B = \{x \in \mathbb{R} : 3 < x \leq 9\},$$

$$C = \{x \in \mathbb{R} : 1 \leq x \leq 6\},$$

$$D = \{x \in \mathbb{R} : -5 < x < 10\},$$

$$E = \{x \in \mathbb{R} : 0 \leq x \leq 11\},$$

$$F = \{x \in \mathbb{R} : -4 < x \leq -3\}.$$

Then $A \cap B \subset C$, $A \cup B \subset D$, $B \cup C \subset E$, and $A \cap E \subset D$. These relationships are illustrated in the Venn diagram in Figure 5.2.

□

5.3.2 Properties of Intersection and Union

There are standard formulas that connect intersection, union, and set-theoretic difference. These include:

(5.3.2.1) $A \setminus B = A \cap {}^c B$;

(5.3.2.2) $A \cap (B \cup C) = (A \cap B) \cup (A \cap C)$;

(5.3.2.3) $A \cup (B \cap C) = (A \cup B) \cap (A \cup C)$;

(5.3.2.4) [De Morgan's Law] ${}^c(A \cup B) = {}^c A \cap {}^c B$;

(5.3.2.5) [De Morgan's Law] ${}^c(A \cap B) = {}^c A \cup {}^c B$;

(5.3.2.6) $A \setminus (B \cup C) = (A \cap {}^c B) \cap (A \cap {}^c C)$;

(5.3.2.7) $A \setminus (B \cap C) = (A \cap {}^c B) \cup (A \cap {}^c C)$.

Of course, the operations of union and intersection are associative:

(5.3.2.8) $A \cap (B \cap C) = (A \cap B) \cap C$;
(5.3.2.9) $A \cup (B \cup C) = (A \cup B) \cup C$.

The reader is encouraged to use Venn diagrams to confirm all of these set-theoretic identities.

5.3.3 Other Set-Theoretic Operations

If $\{S_\alpha\}_{\alpha \in A}$ is a collection of sets, indexed over some (possibly very large) set A, then we let $\cup_\alpha S_\alpha$ denote the union of all the S_α. More precisely,

$$\bigcup_{\alpha \in A} S_\alpha = \{x : x \in S_\alpha \text{ for some } \alpha\}.$$

Likewise, $\cap_\alpha S_\alpha$ is the intersection of all the S_α. More precisely,

$$\bigcap_{\alpha \in A} S_\alpha = \{x : x \in S_\alpha \text{ for all } \alpha\}.$$

If S, T are sets, then $S \times T$ is the set of all ordered pairs of elements of S and T. More precisely,

$$S \times T = \{(s, t) : s \in S, t \in T\}.$$

If $S_1, \ldots, S_k$ are sets, then

$$S_1 \times \cdots \times S_k = \{(s_1, s_2, \ldots, s_k) : s_j \in S_j \ \forall j = 1, \ldots, k\}.$$

If $S_\alpha, \alpha \in A$ are sets indexed over a (possibly very large) set A, then

$$\prod_{\alpha \in A} S_\alpha = \{f : A \to \cup_\alpha S_\alpha | f(\alpha) \in S_\alpha\}.$$

Here f denotes a function (see Section 5.7 for the definition).

If S and T are sets then their *symmetric difference* is defined to be $S \triangle T = (S \setminus T) \cup (T \setminus S)$.

5.4 Relations and Equivalence Relations

5.4.1 What Is a Relation?

A *relation* on sets S and T is a collection of (some but not necessarily all) elements of $S \times T$ (see also Chapter 2 on semantics and syntax).

Example 5.9

Let $S = \{1, 2, 3, 4, 5\}$ and $T = \{a, b, c\}$. Then

$$\mathcal{R} = \{(3, b), (2, a), (1, b), (5, c), (1, c)\}$$

is a relation on S and T.

□

We often find it convenient to denote a relation by the binary symbol $\sim$ (*not* to be confused with the use of this very same symbol to mean "not" in the propositional calculus). Thus, in the previous example, $3 \sim b$, $2 \sim a$, $1 \sim b$, $5 \sim c$, and $1 \sim c$.

Many important relations are of a set S with itself. We call such a relation "a relation on S." If $S = \{x, y, z, w\}$ then

$$\mathcal{R} = \{(x, w), (z, y), (x, z), (w, w)\}$$

is a relation on S.

The set $\operatorname{Dom} \mathcal{R} \equiv \{s : \exists t \text{ such that } (s, t) \in \mathcal{R}\}$ is called the *domain* of the relation. The set $\operatorname{Im} \mathcal{R} \equiv \{t : \exists s \text{ such that } (s, t) \in \mathcal{R}\}$ is called the *image* of the relation. We sometimes call T the *range* or the *codomain* of the relation. In Example 5.9 above, the domain of $\mathcal{R}$ is $\{1, 2, 3, 5\}$, the image is $\{a, b, c\}$, and the range is $\{a, b, c\}$. In the second example, the domain is $\{w, x, z\}$, the image is $\{w, y, z\}$, and the range is $\{x, y, z, w\}$.

5.4.2 Partial and Full Orderings

An important example of a relation is a *partial ordering*. Let $\mathcal{R}$ be a relation on a set S. If $\mathcal{R}$ is reflexive, antisymmetric ($a\mathcal{R}b$ and $b\mathcal{R}a$

implies $a = b$), and transitive (see Definition 5.1 below) then it is called a partial ordering. For instance, the relation $\leq$ on the integers is a partial ordering.

Observe that a partial ordering does not always give rise to a full ordering (defined below). For example, let $S = \mathcal{P}(\mathbb{Z})$. Thus S is the collection of all possible sets of integers. Let the partial ordering be given by $\subseteq$. Then two given subsets A, B of $\mathbb{Z}$ may or may not be comparable under $\subseteq$. If $X \subseteq S$ is a subset with the property that any two elements of X *can* be compared under the partial ordering, then we say that X is *linearly ordered* or *totally ordered.* We call such an ordering a *total ordering* or a *full ordering.*

5.5 Equivalence Relations

5.5.1 What Is an Equivalence Relation?

Certain relations on S are special. For example, let S be the set of all integers. Let $\mathcal{R}$ be the relation "$s \sim t$ if $(s - t)$ is evenly divisible by 2." Then 2 is related to all of the numbers $\ldots - 4, -2, 0, 2, 4, \ldots$. And 1 is related to all of the numbers $\ldots - 3, -1, 1, 3, \ldots$. In fact any integer either is related to all the numbers in the first set (the even integers) or to all the numbers in the second set (the odd numbers). If we let $E = \{\ldots - 4, -2, 0, 2, 4, \ldots\}$ and $O = \{-3, -1, 1, 3\}$ then we see that $E \cap O = \emptyset$ and $S = E \cup O$. Every integer is either related to all the numbers in the first set E or else it is related to all the numbers in the second set O, but not both. This is an example of an "equivalence relation."

Definition 5.1 Let S be a set, and let $\sim$ be a relation on S. If $\sim$ satisfies

[**Reflexive**] $s \sim s$ for every $s \in S$,

[**Symmetric**] if $s \sim t$, then $t \sim s$,

[**Transitive**] if $s \sim t$ and $t \sim u$, then $s \sim u$,

then we say that $\sim$ is an *equivalence relation* on S.

5.5.2 Equivalence Classes

In case $\mathcal{R}$ is an equivalence relation on S, then it is a fundamental fact that $\mathcal{R}$ induces a *partition* of S into disjoint subsets. More precisely, if $s \in S$, then we let

$$[s] = \{t \in S : s \sim t\}.$$

Then the fundamental fact is that if $[s] \cap [s'] \neq \emptyset$, then $[s] = [s']$. The sets $[s]$ are called *equivalence classes* induced by the equivalence relation $\mathcal{R}$. Our fundamental fact says that two equivalence classes are either disjoint or identical.

It follows that the set S is the disjoint union of its equivalence classes. We say that the relation $\mathcal{R}$ *partitions* the set S. If S is a set and $\sim$ an equivalence relation on that set, then we use the notation $S/\sim$ to denote the collection of equivalence classes induced by the equivalence relation.

It is worth noting explicitly that if S is any set and $S = \cup_{\alpha \in A} S_\alpha$ is a partition of S into disjoint subsets, then the partition induces an equivalence relation: $x \sim y$ if and only if x and y lie in the same S_α.

5.5.3 Examples of Equivalence Relations and Classes

Example 5.10

Let S be the set of all integers and define the relation $s \sim t$ if $s - t$ is evenly divisible by 2. Observe that

Reflexive For every s, $s - s$ is evenly divisible by 2.

Symmetric If $s - t$ is evenly divisible by 2, then $t - s$ is evenly divisible by 2.

Transitive If $s - t$ is evenly divisible by 2 and $t - u$ is evenly divisible by 2, then $s - u = (s - t) + (t - u)$ is evenly divisible by 2.

Therefore our relation is reflexive, symmetric, and transitive. We are now guaranteed that S will be partitioned into equivalence classes. These equivalence classes are

$$E = (\text{the set of all integers related to } 2) = (\text{all even integers})$$

$$O = (\text{the set of all integers related to } 1) = (\text{all odd integers}).$$

Finally, $S = E \cup O$.

Of course, this example is the same one that we discussed right before the definition of "equivalence relation."

□

Example 5.11

Let S be the set of all people on Earth. If $s, t \in S$, then say that $s \sim t$ if s and t have at least one parent in common. Is this an equivalence relation?

Let us check the three defining properties of an equivalence relation.

Reflexive Let s be a person. Then s and s have at least one parent in common (in fact, trivially, they have both parents in common), so the relation is reflexive.

Symmetric Let s and t be people. If s and t have a parent in common, then certainly t and s have a parent in common, so the relation is symmetric.

Transitive Let s, t, u be people. Suppose that s and t have a parent in common. Also suppose that t and u have a parent in common. Does it follow that s and u have a parent in common? The answer is "no," because the parent common to s and t may be different from the parent common to t and u, so the relation is *not* transitive.

We find that this is not an equivalence relation, and therefore the set of all people is not thereby partitioned into equivalence classes by this relation. What this observation says, in common English, is that the property of having at least one parent in common does not lead to a well-defined notion of "family." Everyone who comes from a family with divorced parents knows this. As an exercise, check that if we replace "have at least one parent in common" with "have both parents in common," then the relation *does* become an equivalence relation. This simple exercise confirms the notion that the more traditional notion of sibling (having both parents in common) does in fact lead to a true partition, that is, to well-defined families.

□

5.5.4 Construction of the Rational Numbers

Example 5.12

Let S be the set of ordered pairs of integers with second entry not zero. Say that two pairs (p, q) and (p', q') are related if $pq' = p'q$. We verify that this is an equivalence relation:

Reflexive If (p, q) is an ordered pair of integers with $q \neq 0$, then $pq = pq$. Therefore $(p, q) \sim (p, q)$.

Symmetric If $(p, q) \sim (p', q')$, then $pq' = p'q$ and hence $p'q = pq'$ so that $(p', q') \sim (p, q)$.

Transitive] If $(p, q) \sim (r, s)$ and $(r, s) \sim (t, u)$, then we know that

$$ps = qr \tag{$*$}$$

and

$$ru = st. \tag{$**$}$$

Multiplying these two equations together, left side against left side and right side against right, yields

$$psru = qrst. \qquad (***)$$

Now if $r = 0$, then $(*)$ and $(**)$ yield $p = 0$ and $t = 0$. Hence certainly $pu = qt$ and we are done, so we may assume $r \neq 0$, and we may divide both sides of $(***)$ by rs to obtain

$$pu = qt.$$

Thus $(p, q) \sim (t, u)$ and we see that our relation is transitive.

□

In point of fact, the equivalence relation described in the last example is the one used to construct the rational numbers from the integers. Think of the ordered pair (p, q) as being identified with the fraction $\frac{p}{q}$. Then $(p, q) \sim (r, s)$ simply means that $ps = qr$, and that in turn translates to the statement that $\frac{p}{q}$ and $\frac{r}{s}$ represent the same fraction.

5.6 Number Systems

We will say more about the construction of the various number systems in Chapter 7. But we will be using the number systems throughout the book to illustrate various ideas from logic, set theory, and proof theory. So we take a moment now to record, informally, the number systems that are in common use.

Notat.	**Name**	**Brief Description**
$\mathbb{N}$	natural numbers	$\{1, 2, \ldots\}$
$\mathbb{Z}$	integers	$\{\ldots, -2, -1, 0, 1, 2, \ldots\}$
$\mathbb{Q}$	rational numbers	$\{p/q : p, q \in \mathbb{Z}, q \neq 0\}$
$\mathbb{R}$	real numbers	all decimal numbers *or* all limits of rational numbers
$\mathbb{C}$	complex numbers	$\{x + iy : x, y \in \mathbb{R}\}$
$\mathbb{H}$	quaternions	$\{x + \mathbf{i}y + \mathbf{j}z + \mathbf{k}w : x, y, z, w \in \mathbb{R}\}$
$\mathbb{O}$	Cayley numbers	$\mathbb{R}^8$ with a special algebraic structure

It is a fact that only 1-, 2-, 4-, and 8-dimensional space can be equipped with algebraic structures so that the algebraic operations are continuous in the spatial coordinates and so that the resulting number systems have at least some of the essential properties of a field (such as nonzero numbers having multiplicative inverses). In technical language, these number systems are all *division rings*. See [ADA] and [BOM] for all the particulars.

5.7 Functions

It is arguable that the language of sets and functions is the hallmark of twentieth-century mathematics. Now we learn a rigorous definition of the function concept.

5.7.1 What Is a Function?

Definition 5.2 Let $\mathcal{R}$ be a relation on sets X and Y. Suppose that $\mathcal{R}$ satisfies the following two axioms

1. If $x \in X$, then there is a $y \in Y$ such that $(x, y) \in \mathcal{R}$;
2. If $(x, y_1) \in \mathcal{R}$ and $(x, y_2) \in \mathcal{R}$, then $y_1 = y_2$.

Then $\mathcal{R}$ defines a *function* from X to Y. We often denote a function by $f : X \to Y$ or $y = f(x)$.

In common parlance, we often say that a function from X to Y is a "rule" that assigns to each element of X one and only one element of Y. The more formal definition removes any doubt about what the word "rule" means.

The set X in the definition is called the *domain* of the function f. The set Y in the definition is called the *range* or the *codomain* of the function f. The set $I = \{y : \exists x \text{ such that } f(x) = y\}$ is called the *image* of f. [Compare the analogous language for relations in Subsection 5.4.1.] We sometimes denote these sets by $\text{Dom}\, f$, $\text{Ra}\, f$, and $\text{Im}\, f$ respectively. The letters f, g, h are commonly used to denote functions.

In case a function f has as its domain only a (possibly proper) *subset* of X, then we say that f is a *partial function* from X to Y.

5.7.2 Examples of Functions

Example 5.13

Let $X = \{1, 2, 3, 4, 5\}$ and $Y = \{a, b, c, d\}$. Let

$$\mathcal{R} = \{(1, c), (2, a), (3, c), (4, a), (5, d)\}.$$

Then, by inspection, $\mathcal{R}$ is a function from X to Y. We would more commonly write

$$\begin{aligned} f(1) &= c, \\ f(2) &= a, \\ f(3) &= c, \\ f(4) &= a, \\ f(5) &= d. \end{aligned}$$

We see that each element of X has one and only one element of Y assigned to it.

□

Example 5.14

Let $X = \{1, 2, 3, 4, 5\}$ and $Y = \{a, b, c, d\}$. Let

$$\mathcal{R} = \{(1, c), (2, a), (3, c), (4, a), (1, d), (5, d)\}.$$

Then $\mathcal{R}$ is *not* a function because both of the range elements c and d are assigned to the domain element 1.

Let

$$\mathcal{S} = \{(1, c), (2, a), (3, c), (5, d)\}.$$

Then $\mathcal{S}$ is *not* a function because the domain element 4 has no element of Y assigned to it.

□

5.7.3 One-to-One or Univalent

If a function $f : X \to Y$ has the property that $f(x_1) = f(x_2)$ implies $x_1 = x_2$, then we say that f is one-to-one. In the language of ordered pairs, if (x_1, y) and (x_2, y) are both elements of the function, then we demand that $x_1 = x_2$. In common parlance, a function f is one-to-one if different elements of the domain are mapped by f to different elements of the image. A one-to-one function is sometimes called *univalent* or *injective*.

Example 5.15

Let $f : \mathbb{R} \to \mathbb{R}$ be given by $f(x) = x^2$. Then f is certainly a function, but f is not one-to-one. In fact $f(-1) = f(1)$.

□

Example 5.16

Let $g : \mathbb{R} \to \mathbb{R}$ be given by $g(x) = 3x + 2$. Then g is a function, and g is one-to-one. To see this, observe that if $g(x_1) = g(x_2)$, then $3x_1 + 2 = 3x_2 + 2$, and hence (solving) $x_1 = x_2$.

□

Example 5.17

Return to the function $f : \mathbb{R} \to \mathbb{R}$, $f(x) = x^2$ from Example 5.15. We may restrict the domain of f by insisting that we only consider values of x that lie in $A \equiv \{x : 0 \leq x < +\infty\}$. Then $f : A \to \mathbb{R}$ is now a one-to-one function. For if $f(x_1) = f(x_2)$ and $x_1, x_2 \geq 0$, then it must be that $x_1 = x_2$ (i.e., any nonnegative number has but one nonnegative square root).

□

5.7.4 *Onto or Surjective*

If a function $f : X \to Y$ has the property that $y \in Y$ implies that there exists an $x \in X$ such that $f(x) = y$, then we say that f is *onto* (or *onto* Y). In the language of ordered pairs: if $y \in Y$ then there is an $x \in X$ such that (x, y) is an element of the function. In common parlance, a function f is onto if each element of Y is in the image of f. An onto function is sometimes called *surjective.* A function is surjective if its image equals its range.

Example 5.18

Let $f : \mathbb{R} \to \mathbb{R}$ be given by $f(x) = x^2$. Then f is certainly a function, but f is not onto. In fact, the point $-3 \in Y \equiv \mathbb{R}$ is not in the image of f. Of course we may modify the definition of f by letting $B \equiv \{y \in \mathbb{R} : y \geq 0\}$. Then we think of $f : \mathbb{R} \to B$, $f(x) = x^2$. With this modified definition (i.e., with altered range), the function f is onto.

□

Example 5.19

Let $g : \mathbb{R} \to \mathbb{R}$ be given by $g(x) = 3x + 2$. Then g is a function, and g is onto. For if $y \in Y \equiv \mathbb{R}$, then $x = [y - 2]/3$ is mapped by g to y. □

Example 5.20

There is no logical relationship between the properties of "one-to-one" and "onto." To wit:

- The function $g : \mathbb{R} \to \mathbb{R}$ given by $g(x) = x^3$ is both one-to-one and onto.
- The function $f : \mathbb{R} \to \{y \in \mathbb{R} : y \geq 0\}$ given by $f(x) = x^2$ is onto but not one-to-one.
- The function $h : \mathbb{R} \to \mathbb{R}$ given by $h(x) = \exp x$ is one-to-one but not onto.
- The function $m : \mathbb{R} \to \mathbb{R}$ given by $m(x) = \log(|x| + 2)$ is neither one-to-one nor onto.

□

5.7.5 Set-Theoretic Isomorphisms

Let S and T be sets. If $f : S \to T$ is a function that is both one-to-one and onto, then we call f a *set-theoretic isomorphism.* In a certain sense, the existence of a set-theoretic isomorphism between S and T means that S and T have the same number of elements (when S and T are finite sets this assertion is obvious). This idea will be developed in greater detail in the next section.

5.8 Cardinal Numbers

5.8.1 Comparison of the Sizes of Sets

What does it mean to say that two sets have the same size, or the same number of elements? How can we compare two sets?

A useful way to answer these questions is as follows. We say that two sets S and T have the *same cardinality* if there is a function $f : S \to T$ such that f is one-to-one and onto. We call such a function f a "set-theoretic equivalence" or a "set-theoretic isomorphism" or a "one-to-one correspondence." We will write $\text{card}(S) = \text{card}(T)$.

Example 5.21

Let $S = \{1, 2, 3, 4, 5, 6\}$ and $T = \{A, B, C, D, E, F\}$. The function

$$f : \begin{cases} 1 \longmapsto A \\ 2 \longmapsto B \\ 3 \longmapsto C \\ 4 \longmapsto D \\ 5 \longmapsto E \\ 6 \longmapsto F \end{cases}$$

is both one-to-one and onto. Thus S and T have the same cardinality; that is, $\text{card}(S) = \text{card}(T)$.

□

5.8.2 *Cardinality and Cardinal Numbers*

In the preceding example, the notion of "same cardinality" makes rigorous the notion that S and T have the same number of elements. For finite sets, the concept does not really tell us anything new. But for infinite sets the idea (due to Georg Cantor, 1845–1918) is profound. In case a set S is finite, we use the notation $|S|$ to denote the number of elements in S, and we call $|S|$ the "cardinality" of the set S.

Example 5.22

Let $S = \mathbb{Z}$ (the integers) and $T = 2\mathbb{Z}$ = the even integers. Then we claim that S and T have the same cardinality.

In fact, the required function f is

$$\begin{aligned} f : S &\to T \\ n &\mapsto 2n. \end{aligned}$$

It is immediate that f takes integers to even integers, that it is one-to-one, and that it is onto. Hence S and T have the same cardinality.

□

What is remarkable about the last example is that T is a proper subset of S, yet we are asserting that, in the sense of cardinality, S and T have the same number of elements. Such an equivalence is not possible for finite sets S and T.

Example 5.23

Let $S = \mathbb{Q}^p$, the set of positive rational numbers. And let $T = \mathbb{N}$, the set of natural numbers (or positive integers). Then S and T have the same cardinality.

$$\begin{array}{ccccc}
\frac{1}{1} & \frac{1}{2} & \frac{1}{3} & \frac{1}{4} & \cdots \\
\frac{2}{1} & \frac{2}{2} & \frac{2}{3} & \frac{2}{4} & \cdots \\
\frac{3}{1} & \frac{3}{2} & \frac{3}{3} & \frac{3}{4} & \cdots \\
\frac{4}{1} & \frac{4}{2} & \frac{4}{3} & \frac{4}{4} & \cdots \\
\cdots & & \cdots & &
\end{array}$$

Figure 5.3

To see this, we lay out the rational numbers in a tableau (Figure 5.3).

We now associate positive integers in a one-to-one fashion with the numbers in this tableau. We do so by beginning in the upper-left-hand corner and then proceeding along diagonals stretching from the lower left to the upper right (Figure 5.4).

This scheme clearly associates one positive integer to each fraction, and the association is one-to-one and onto:

$$\begin{array}{ccccccccccc}
1 & 2 & 3 & 4 & 5 & 6 & 7 & 8 & 9 & 10 & \cdots \\
\frac{1}{1} & \frac{2}{1} & \frac{1}{2} & \frac{3}{1} & \frac{2}{2} & \frac{1}{3} & \frac{4}{1} & \frac{3}{2} & \frac{2}{3} & \frac{1}{4} & \cdots
\end{array}$$

Note that every fraction is counted multiple times, because the fraction $\frac{1}{2}$ also appears as $\frac{2}{4}$ and $\frac{3}{6}$ and so on. But we can skip the repeats, and the counting scheme still works.

□

The preceding example is even more startling than the one before, because the positive integers form (apparently) a quite small subset of the positive rational numbers. Yet we are showing that the two sets have precisely the same number of elements. And we do so in a very graphic manner, exhibiting the correspondence quite explicitly.

5.8.3 An Uncountable Set

It is natural to wonder whether every infinite set can be placed in one-to-one correspondence with the integers, or with the positive integers. The answer is "no." In fact, the set of all sequences of 0's and 1's forms a strictly larger infinite set, as the next example shows.

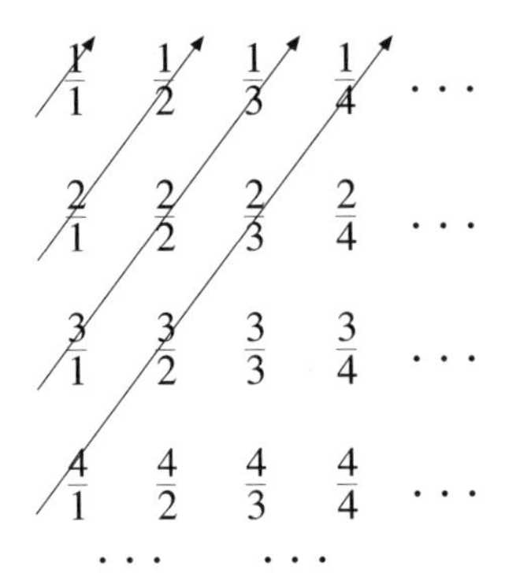

Figure 5.4

Example 5.24

Let S be the set of all sequences of 0's and 1's. We claim that S does *not* have the same cardinality at $\mathbb{N}$, the set of all positive integers.

The argument is by contradiction. Suppose, to the contrary, that there is a one-to-one correspondence between the set T of all positive integers and the set S of all sequences of 0's and 1's. Thus we can make a list:

$$\begin{array}{lccccc} 1. & a_1^1 & a_2^1 & a_3^1 & a_4^1 & \ldots \\ 2. & a_1^2 & a_2^2 & a_3^2 & a_4^2 & \ldots \\ 3. & a_1^3 & a_2^3 & a_3^3 & a_4^3 & \ldots \\ 4. & a_1^4 & a_2^4 & a_3^4 & a_4^4 & \ldots \\ \ldots & & \ldots & & \ldots & \end{array}$$

Here each a_j^k is either a 0 or a 1. Thus the first row is the sequence of 0's and 1's (the element of S) corresponding to **1**; the second row is the sequence of 0's and 1's (the element of S) corresponding to **2**; the third row is the sequence of 0's and 1's (the element of S) corresponding to **3**; and so forth. We claim to have explicitly exhibited a one-to-one correspondence between the set of positive integers and the collection S of all sequences of 0's and 1's.

But now we will find that this enumeration is in error. In fact, no matter how cleverly we think we have enumerated all the elements of S, there will always be a sequence of 0's and 1's that has been omitted from the list. That sequence is:

The first element is 0 if a_1^1 is 1 and 1 if a_1^1 is 0.
The second element is 0 if a_2^2 is 1 and 1 if a_2^2 is 0.
The third element is 0 if a_3^3 is 1 and 1 if a_3^3 is 0.
The fourth element is 0 if a_4^4 is 1 and 1 if a_4^4 is 0.

...And so forth ...

In other words, we are constructing a new sequence that differs from the first in the list in the first entry, differs from the second in the list in the second entry, differs from the third in the list in the third entry, and so forth. Certainly the sequence we have now constructed cannot be on the list, so our claim to have enumerated all sequences of 0's and 1's cannot be true. That is a contradiction.

We conclude that the collection of all sequences of 0's and 1's cannot be enumerated.

□

5.8.4 Countable and Uncountable

If a set S has the same cardinality as the set $\mathbb{N}$ of positive integers, then we say that S is *countable*. Thus it is immediate that the set $\mathbb{N}$ of positive integers is countable. A simple argument (similar to our first example) shows that the set of all integers is countable, the set of even integers is countable, and the set of odd integers is countable. The set of positive rational numbers is also countable. Example 5.24 shows that the set of sequences of 0's and 1's is not countable. It also follows that the set of real numbers is not countable, for any real number has a unique binary expansion (analogous to a decimal expansion but in base 2). And that is nothing other than a sequence of 0's and 1's.

If a set is infinite but is not countable, then we say it is *uncountable*.

Example 5.25

Let S be the set of all subsets of the positive integers. Then S is uncountable.

To see this, observe that we can associate to any subset of the positive integers a sequence of 0's and 1's. We do so as follows. Let X be such a subset. If $1 \in X$, then the first element of the associated sequence is 1, otherwise it is 0. If $2 \in X$, then the second element of the associated sequence is 1, otherwise it is 0. If $3 \in X$, then the third element of the associated sequence is 1, otherwise it is 0, and so forth. In this way, every subset $X \subset S$ has associated to it a sequence of 0's and 1's (what amounts to the "indicator function" of the set—see our discussion in Subsection 5.10.1) and vice versa. Since the set of all such sequences is uncountable, then so is the set of subsets of the positive integers uncountable.

□

5.8.5 Comparison of Cardinalities

In fact the preceding example is an instance of a very general phenomenon that is well worth recording explicitly. Before we do so, let us introduce a convenient piece of terminology.

Definition 5.3 Let S and T be sets. We say that T has *greater cardinality* than S if there is a one-to-one function $f : S \to T$ but there is no one-to-one function $g : T \to S$. We sometimes write $\text{card}(S) < \text{card}(T)$.

The appropriateness of this last definition is made clear by the following result.

Theorem 5.1 (Schroeder–Bernstein)
Let S and T be sets. If there exists a one-to-one function $f : S \to T$ and a one-to-one function $g : T \to S$, then S and T have the same cardinality. In other words, the conclusion is that there exists a function $h : S \to T$ that is one-to-one and onto.

Certainly there is a one-to-one function from the natural numbers $\mathbb{N}$ to the collection T of all sequences of 0's and 1's. Namely, associate to the positive integer k the sequence consisting of all 0's except for a 1 in the kth place. But Example 5.24 shows that there is no one-to-one function from T to $\mathbb{N}$, so the set T has greater cardinality than the cardinality of $\mathbb{N}$.

5.8.6 The Power Set

Let S be any set. Then its *power set* $\mathcal{P}(S)$ is the collection of all subsets of S. For example, if $S = \{1, 2, 3\}$ then

$$\mathcal{P}(S) = \Big\{\emptyset, \{1\}, \{2\}, \{3\}, \{1,2\}, \{2,3\}, \{1,3\}, \{1,2,3\}\Big\}.$$

Notice that we include the empty set $\emptyset$ in the enumeration of all subsets of S. The empty set is a subset of *any* set. In general, if S is a finite set with k elements, then $\mathcal{P}(S)$ will have 2^k elements (Section 8.2.2 contains a proof of this assertion). As a result, some books denote the power set of S by 2^S. If S is infinite, then of course the power set of S will be infinite. An important theorem of basic set theory is as follows.

Theorem 5.2 (Cantor)
Let S be any set. Then the cardinality of $\mathcal{P}(S)$ is strictly greater than the cardinality of S.

One consequence of this result is that there are sets of arbitrarily large cardinality. For, given any set S, $\mathcal{P}(S)$ will have cardinality greater than the cardinality of S itself.

5.8.7 The Continuum Hypothesis

It was an open problem for many years to determine whether there is a set S, $\mathbb{N} \subset S \subset \mathbb{R}$, such that S has greater cardinality than $\mathbb{N}$ and lesser cardinality than $\mathbb{R}$. In classical language, we let $\aleph_0$ denote the cardinality of $\mathbb{N}$ and c denote the cardinality of $\mathcal{P}(\mathbb{N})$ (equivalently, the cardinality of $\mathbb{R}$). The question, then, is whether there is a cardinal number $\mathbf{m}$ such that $\aleph_0 < \mathbf{m} < c$. To carry the notation a bit further: we let $\aleph_1$ denote the least cardinal that exceeds $\aleph_0$ (such a cardinal may be unambiguously defined—see [COH, p. 67]). Is it true that $c = \aleph_1$?

Cantor believed that $\aleph_1 < c$, and the problem itself came to be called the *continuum hypothesis* (*CH*). In 1961, Paul Cohen proved that the continuum is independent of the other axioms of set theory. In other words, it is consistent with the axioms to suppose that the continuum hypothesis is false. (It had been known for some time—see below—that it is consistent with the axioms of set theory to suppose that *CH* is true.) The *generalized continuum hypothesis* (*GCH*) asserts that if S is a set and $\mathcal{P}(S)$ is its power set, then there is no set T with cardinality strictly between the cardinalities of S and $\mathcal{P}(S)$. In more formal language, the generalized continuum hypothesis is that $\text{card}(\mathcal{P}(\aleph_n)) = \text{card}(\aleph_{n+1})$.[1] Kurt Gödel had already proved that it was consistent with the other axioms to suppose that the continuum hypothesis is true, and also that the generalized continuum hypothesis is true.

What this all means is that the truth or falsity of the continuum hypothesis cannot be proved from the axioms of set theory. The situation (from the point of view of logic) is quite similar to that with the parallel postulate in Euclidean geometry. It is consistent with the first four axioms of Euclid to suppose that the parallel postulate is true. It is also consistent with the other axioms of Euclid to suppose that the parallel postulate is false. There are geometries in which the parallel postulate holds; and there are geometries in which the parallel postulate fails. Just so, there are models of set theory in which the continuum

[1] Here we indulge in the common abbreviation of letting $\aleph_n$ denote some set with cardinality $\aleph_n$ and likewise for $\aleph_{n+1}$—see [COH, p. 67].

hypothesis is true; and there are models of set theory in which the continuum hypothesis is false.

5.8.8 Martin's Axiom

Martin's Axiom (MA) is an axiom for set theory that has been proposed as an alternative to the continuum hypothesis. Martin and Solovay [MAS] observed that, in many application of the continuum hypothesis, the only properties of *CH* that we actually used were those encapsulated in MA. To formulate Martin's Axiom, we need a few definitions.

Let $(P, \leq)$ be a partially ordered set. We say that $D \subseteq P$ is *dense* if, for each $p \in P$, there is a $d \in D$ with $d \leq p$. A subset $Q \subseteq P$ is *compatible* provided that, for each finite set $F \subseteq Q$, there is a $q \in P$ such that $q \leq p$ for all $p \in F$. Finally, the partially ordered set $(P, \leq)$ is said to satisfy the *countable chain condition* (ccc) if every pairwise incompatible (i.e., no two elements can be compared using $\leq$) subset is countable.

> **Martin's Axiom** Suppose that $(P, \leq)$ is a ccc partially ordered set and that $\mathcal{D}$ is a family of less than 2^ω dense subsets of $\mathbf{P}$ (where ω is the first infinite cardinal, or the cardinality of the natural numbers). Then there is a compatible subset Q of $\mathbf{P}$ that meets every member of $\mathcal{D}$.

Martin's Axiom was discovered in the context of Souslin trees. The relationship of this new axiom to other parts of set theory has been rather thoroughly explored. We list just a few of the results here, and refer the reader to the delightful article of M. E. Rudin in [BAR] for further details and references. For this discussion, recall that ZFC is Zermelo–Fraenkel set theory *with* the Axiom of Choice.

- Martin's Axiom is independent of $\sim CH$.
- If ZFC is consistent, then $ZFC + MA + \sim CH$ is consistent.
- The continuum hypothesis implies MA.
- If we assume MA and $\sim CH$, then there is no Souslin tree.
- Assume Martin's Axiom. If X is a ccc, compact Hausdorff space, then X is not the union of fewer than 2^ω nowhere dense sets.
- Assume Martin's Axiom. If $\omega \leq \lambda < 2^\omega$, then $2^\omega = 2^\lambda$.
- Assume Martin's Axiom. Let m be Lebesgue measure. If $0 \leq \lambda < 2^\omega$ and if, for each $\alpha \in \lambda$, $X_\alpha \subseteq \mathbb{R}$ and $m(X_\alpha) = 0$, then $m(\cup_{\alpha \in A} X_\alpha) = 0$.

- Assume Martin's Axiom. If $\mathcal{B}$ is a family of cardinality less than 2^ω of subsets of ω with each finite subset of $\mathcal{B}$ having infinite intersection, then there is an infinite subset L of ω such that $L \setminus B$ is finite for all $B \in \mathcal{B}$.

5.8.9 Inaccessible Cardinals and Measurable Cardinals

Let A and B be cardinals; that is, A is an equivalence class of sets, each having the same cardinality (and hence set-theoretically isomorphic to each other) and similarly for B. We say that $A < B$ if, given $a \in A$ and $b \in B$, there is an injection $\varphi : a \to b$ but there exists no injection $\psi : b \to a$.

Now we can formulate the Axiom of Inaccessible Cardinals:

Axiom: There is an uncountable cardinal A such that

1. If $B < A$, then A is not the sum (union) of B cardinals each of which is $< A$.
2. If $B < A$, then the cardinality of $\mathcal{P}(B)$ is also $< A$.

The Axiom of Inaccessible Cardinals is independent of the usual axioms of *ZF*. It is used, for instance, to construct new models of set theory. As an example, if C is an inaccessible cardinal, then the set of all sets of cardinality less than C is a model for *ZF*.

If there is a set S equipped with a nontrivial, countably additive, finite, real-valued measure defined on *all* subsets of S and such that the measure of each singleton is zero, then we say that S is a *measurable cardinal.* It is known, of course, that the real line equipped with Lebesgue measure is *not* a measurable cardinal. If one assumes the generalized continuum hypothesis, then it can be shown that the cardinality of a measurable cardinal S must exceed the first inaccessible cardinal. It is not known whether measurable cardinals exist (in such a way as to be consistent with the usual axioms of *ZF*).

5.8.10 Ordinal Numbers

The ordinal numbers are a generalization of the natural numbers. The construction of the ordinal numbers is, in effect, a formalization of the concept of mathematical induction. In particular, it makes possible the method of *transfinite induction.* A detailed treatment may be found in [SUP2, Chapter 7]. We now consider these ideas.

Intuitively, an ordinal number represents an equivalence class of well-ordered sets. Rigorously, we proceed as follows:

Definition 5.4 A relation $\mathcal{R}$ on a set S is said to be an *order* (or an *order relation*) on S if

1. For all $x, y \in S$, one and only one of the relations $x\mathcal{R}y$, $y\mathcal{R}x$, $x = y$ holds.
2. If $x, y, z \in S$, $x\mathcal{R}y$, and $y\mathcal{R}z$ then $x\mathcal{R}z$.

We generally write an order relation as $x < y$ rather than $x\mathcal{R}y$.

Definition 5.5 An order relation $<$ on S is said to *well-order* S if each $B \subseteq S$ has a least element: there exists a $b_0 \in B$ such that $b \not< b_0$ for all $b \in B$.

The standard ordering on the natural numbers $\mathbb{N}$ well-orders $\mathbb{N}$. In fact, the number 1 is the least element of $\mathbb{N}$, and each subset of $\mathbb{N}$ has a first element that is also the least element. The standard ordering on the rational numbers $\mathbb{Q}$ does *not* well-order $\mathbb{Q}$. It can be proved that any set can be well-ordered (with *some* ordering); this assertion is equivalent to the Axiom of Choice.

Definition 5.6 An *ordinal* is a set a that is well-ordered by the relation $\in$ and is transitive. Here we say that a set a is transitive if $b \in a$ and $c \in b$ implies $c \in a$.

We now list some basic facts about ordinals:

- The sets $\emptyset, \{\emptyset\}, \{\emptyset, \{\emptyset\}\}, \ldots$ are each ordinals. In general, if $n+1 = n \cup \{n\}$, then the sets n are ordinals. These form a model for the natural numbers.
- Any well-ordered set is isomorphic to some ordinal.
- Any ordinal is equal to the set of all ordinals that precede it.
- If a is an ordinal, then $b = a \cup \{a\}$ is the least ordinal that is greater than a.
- If S is a set of ordinals, then there is a least ordinal in S.
- If S is a set of ordinals, then there is a least ordinal a such that $b \in S$ implies $b < a$. We write $a = \sup S$.

We say that an ordinal a is a *successor* if there is an ordinal b such that $a = b + 1 \equiv \{c : c \leq b\}$. We say that a is a *limit ordinal* if $a \neq 0$ and a is not a successor.

If a is an ordinal and if $b \leq a$ implies that b is a successor or 0, then a is an integer.

5.8.11 Mathematical Induction

Mathematical induction is an important proof method throughout mathematics. Here we treat the rubric formally. See also the treatment in Section 8.2.

Theorem 5.3 (mathematical induction)
Suppose that ω is the first countable ordinal (we have also used this symbol to denote the first infinite cardinal). Assume that

- $x \subseteq \omega$;
- $\emptyset \in x$;
- $n \in x \Rightarrow n + 1 \in x$.

Then $x = \omega$.

See Sections 5.8.11, 5.8.12, and 8.2 for more on induction.

5.8.12 Transfinite Induction

See [COH, p. 63] for this formulation of transfinite induction. The book [SUP2] also has an elegant treatment.

Theorem 5.4 (transfinite induction)
Let A be a formula. Let t_i be given and suppose that for all x there is a unique y such that $A(x, y; t_1, \ldots, t_k)$. Thus A defines a function $y = \varphi(x)$. Then, for every ordinal a and set z, there is a unique function f defined on $\{b : b \leq a\} \equiv a + 1$ so that $f(0) = z$ and, for all $b \leq a$, it holds that $f(b) = \varphi(h)$, where h is the function f restricted to b.

If x and y are ordinal numbers, then we write

$$x <_o y \text{ to mean that } x, y \text{ are ordinals and } x \in y$$

and

$$x \leq_o y \text{ to mean that } x, y \text{ are ordinals and } [x = y \text{ or } x <_o y].$$

We write $\text{Lim}(\alpha)$ to mean that α is a limit ordinal.

Now we can give a second formulation of transfinite induction:

Theorem 5.5 (transfinite induction, second version)

(a) Let X be a set. Then

$$[0 \in X \wedge \forall \alpha, (\alpha \in X \Rightarrow \alpha' \in X) \wedge$$

$$\forall \alpha, (\mathrm{Lim}(\alpha) \wedge \forall \beta, (\beta <_o \alpha \Rightarrow \beta \in X) \Rightarrow \alpha \in X)] \Rightarrow$$

all ordinals lie in X.

(b) Induction up to the ordinal number δ:

$$[0 \in X \wedge \forall \alpha, (\alpha <_o \delta \wedge \alpha \in X \Rightarrow \alpha' \in X) \wedge \forall \alpha, (\alpha <_o \delta \\ \wedge \mathrm{Lim}(\alpha) \wedge \forall \beta, (\beta <_o \alpha \Rightarrow \beta \in X) \Rightarrow \alpha \in X)] \Rightarrow \delta \subset X.$$

5.9 A Word About Classes

5.9.1 Russell's Paradox

First we consider Russell's paradox. Let

$$S = \{X : X \text{ is a set and } X \notin X\}.$$

To see that S is not *a priori* preposterous, note that

$$Y = \{X : X \text{ is a set that can be described} \\ \text{with fewer than fifty words}\}$$

is in fact an element of itself. Most of the sets that we usually encounter are in fact *not* elements of themselves. The question now is, "Is S an element of itself?" If in fact $S \in S$, then, by its very definition, $S \notin S$. If instead $S \notin S$, then (again by its very definition) $S \in S$. Thus we have an absurd situation: S can neither be an element of itself nor can it not be an element of itself.

There are a number of ways to explain away this paradox. The simplest is to note that when we are forming a set z by choosing its members, we do not yet have the object z, so we cannot use it as a member of z. In other words, the set z can have as members only those sets or objects that are formed or defined *before* z. Observe that the Axiom of Regularity certainly forbids sets such as S.

5.9.2 The Idea of a Class

Another way to think about Russell's paradox is this: It turns out that set theory cannot allow the formulation of sets that are too large. The set S above is in fact unallowably large. The theory of classes was developed to describe very large collections of objects. Generally speaking a *class* will be a collection of sets. The definition of a class $\mathcal{C}$ will have the form

$$\mathcal{C} = \{X : \varphi(X)\},$$

where each variable of φ other than X denotes a set. A concise discussion of classes appears in [COH, p. 74] or [BAR, p. 339].

It should be observed that no formula of the language of set theory says anything about all classes. In each instance, a version for classes must be established separately. The tricky part about the theory of classes is establishing the existence of a class.

For the purposes of this handbook, set theory will suffice. We will say nothing more about classes.

5.10 Fuzzy Set Theory

5.10.1 Introductory Remarks About Fuzzy Sets

Fuzzy set theory is an extension of classical set theory that is designed to provide a calculus for considering imprecision, particularly with regard to membership in a set. Fuzzy set theory rejects the traditional "law of the excluded middle" of Aristotle. In classical set theory, if X is a universal set and $S \subseteq X$, then we define the *characteristic function* (or *indicator function*) of S to be

$$\chi_S(x) = \begin{cases} 0 \text{ if } x \notin S \\ 1 \text{ if } x \in S. \end{cases}$$

Thus membership in S is a classical "0–1 game." In fuzzy set theory, we allow the characteristic function to take all values between 0 and 1 inclusive. A value of the characteristic function that is positive but less than one indicates a degree of assuredness with which we can say that the indicated element is in S, and a value equal to one says that the element most certainly is in S.

In fact, fuzzy set theory is used in the design of computer hard drives and has other applications to expert systems, to neural networks, and to process modeling, to name just three. In this brief section, we will outline some of the basic tenets of fuzzy set theory. We will say nothing of the applications.

5.10.2 Fuzzy Sets and Fuzzy Points

Fix once and for all a nonempty set X to be the universe. Let $I = [0, 1]$ be the closed unit interval. A fuzzy set is a pair (X, A), where A is a function: $A : X \to I$. We call A the *membership function* (compare the characteristic function described above).

The value $A(x)$ is the *membership degree* of x to A. By a *fuzzy set*, we mean a function A as in the preceding paragraph. In some sense, $A(x)$ measures the plausibility of the statement "x belongs to A." If $A(x) = 0$, then x *definitely does not* belong to A. If $A(x) = 1$, then x *definitely does* belong to A. A fuzzy set A is called *crisp* or *nonambiguous* if $A(x)$ only takes the values 0 and 1. Thus the (classical) characteristic function of a set is nonambiguous. The empty set $\emptyset$ is the function that is identically 0. We still denote it by $\emptyset$.

If A is a fuzzy set, then the *complement* of A is the fuzzy set $\overline{A} = 1 - A$. The complement of the empty set is the characteristic function of the universal set X, or χ_X.

Two fuzzy sets A and B are equal if $A(x) = B(x)$ for all $x \in X$. We say that

$$A \subseteq B \quad \text{iff} \quad A(x) \leq B(x) \; \forall x \in X.$$

The *product* of fuzzy sets A and B is the function $A \cdot B$ (in some classical set theory books, the intersection of two sets is called their "product"). The *difference* of fuzzy sets A and B is the fuzzy set $A - B$ defined by

$$(A - B)(x) = \max\{A(x) - B(x), 0\} \quad \forall x \in X.$$

A *fuzzy point* in X is a fuzzy set that takes the value zero at all points in X except one. If A is a fuzzy set, then we let the *support* of A be

$$\operatorname{supp} A = \{x \in X : A(x) > 0\}.$$

Now suppose that A is a fuzzy point and that the support of A is $\{y\}$. Let $b = A(y)$. Then we set

$$f_y^b(x) = \begin{cases} b \text{ if } x = y \\ 0 \text{ if } x \neq y. \end{cases}$$

The fuzzy point f_y^b is called *crisp* if $b = 1$.

A fuzzy point f_y^b *belongs* to a fuzzy set A if $f_y^b(x) \leq A(x)$ for all x. We write $f_y^b \in A$. It follows that $A \subseteq B$ if and only if $[f_x^a \in A \Rightarrow f_x^a \in B]$. Also $A = B$ if and only if $[f_x^a \in A \iff f_x^a \in B]$.

5.10.3 An Axiomatic Theory of Operations on Fuzzy Sets

We define two functions $F, G : I \times I \to I$ by the conditions

$$(A \cap B)(x) = F(A(x), B(x)) \quad \forall x \in X,$$

$$(A \cup B)(x) = G(A(x), B(x)) \quad \forall x \in X.$$

The idea is that we will use F and G to specify what intersection and union of fuzzy sets must mean.

To determine what F and G must therefore be, we note that we would certainly want these functions, and their corresponding operations, to satisfy certain fundamental properties, as follows.

(a) The operations induced by F and G must reduce to the classical operations of intersection and union when F and G are crisp (i.e., represent classical sets).

(b) The fuzzy set operations must satisfy the "extremal" conditions

$$A \cap X = A,$$

$$A \cap \emptyset = \emptyset,$$

$$A \cup X = X,$$

$$A \cup \emptyset = A.$$

(c) The set operations must be associative and commutative. The functions F and G must be monotone in each variable.

(d) The functions F and G must satisfy the De Morgan laws:

$$\overline{A \cap B} = \overline{A} \cup \overline{B}$$

and

$$\overline{A \cup B} = \overline{A} \cap \overline{B}.$$

It follows that F and G must satisfy either of the equivalent conditions

$$F(a, b) = 1 - G(1 - a, 1 - b)$$

or

$$G(a, b) = 1 - F(1 - a, 1 - b).$$

We impose the following axioms on the fuzzy set operators F and G:

Axiom 1 The functions F and G take these values:

$$\begin{aligned} F(1,1) &= 1. \\ F(0,0) &= F(0,1) \\ &= F(1,0) \\ &= 0. \end{aligned}$$

$$\begin{aligned} G(0,0) &= 0. \\ G(0,1) &= G(1,0) \\ &= G(1,1) \\ &= 1. \end{aligned}$$

Axiom 2 The functions F and G take these values:

$$F(a, 1) = a.$$
$$F(a, 0) = 0, \forall a \in I.$$

$$G(a, 1) = 1.$$
$$G(a, 0) = a, \forall a \in I.$$

Axiom 3 (Commutativity)

$$F(a, b) = F(b, a), \forall a, b \in I.$$
$$G(a, b) = G(b, a), \forall a, b \in I.$$

Axiom 4 (Associativity)

$$F(F(a, b), c) = F(a, F(b, c)), \forall a, b, c \in I.$$
$$G(G(a, b), c) = G(a, G(b, c)), \forall a, b, c \in I.$$

Axiom 5 (Monotonicity)

$$a \leq a', \ b \leq b' \Rightarrow F(a, b) \leq F(a', b'),$$
$$a \leq a', \ b \leq b' \Rightarrow G(a, b) \leq G(a', b'),$$

for all $a, b, a', b' \in I$.

Axiom 6 (De Morgan's Law)

$$F(a, b) = 1 - G(1 - a, 1 - b), \forall a, b \in I.$$

5.10.4 Triangular Norms and Conorms

These axioms lead to the definition of triangular norms (t-norms). Here a function $T : I \times I \to I$ is a t-norm if

(i) $T(a, 1) = a \quad \forall a \in I$;

(ii) $T(a, b) \leq T(u, v)$ if $a \leq u, \ b \leq v$;

(iii) $T(a, b) = T(b, a)$;

(iv) $T(T(a, b), c) = T(a, T(b, c))$.

The function $S(a,b) = 1 - T(1-a, 1-b)$ is then called a *triangular conorm* or *t-conorm*.

Some of the most standard examples of t-norms and accompanying t-conorms are

$$T_0(x,y) = \min(x,y),$$
$$S_0(x,y) = \max(x,y),$$

$$T_1(x,y) = xy,$$
$$S_1(x,y) = x + y - xy,$$

$$T_\infty(x,y) = \max(x+y-1, 0),$$
$$S_\infty(x,y) = \min(x+y, 1),$$

$$T_s(x,y) = \log_s\left(1 + \frac{(s^x - 1)(s^y - 1)}{s-1}\right), s > 0, s \neq 1,$$
$$S_s(x,y) = 1 - \log_s\left(1 + \frac{(s^{1-x} - 1)(s^{1-y} - 1)}{s-1}\right), s > 0, s \neq 1.$$

At a basic level, understanding relationships among fuzzy sets amounts to studying functional equations for the t-norms and t-conorms. Franks's theorem [DLJ, p. 229] gives a prime example.

Fuzzy sets and fuzzy logic are used to study clustering and classification problems in AI systems and neural networks. They have become a standard part of the language of certain types of engineering.

5.11 The Lambda Calculus

The premise of the Lambda calculus is that the names of functions can be suppressed in favor of focusing on what the function actually *does*. The Lambda calculus is an outgrowth of ideas of Schönfinkel, as developed by Curry and others. Here we can only indicate the basic notions of the Lambda theory, and refer the reader to [NIS], [BAR], and [CF] for further details.

Lambda-functions have two possible syntaxes:

- λ (variable) • (function_body)

or

- λ (signature) | (constraint) • (function_body)

In this brief exposition we concentrate on the first of these; we call this the *pure λ-calculus*.

The formal rules for the λ-calculus are as follows.

λ-Abstraction Syntax

(1) All variables and constants are λ-terms. These entities are called *atoms.*

(2) If M and N are λ-terms, then $(M\ N)$ is also a λ-term. With reference to a term of the form $(M\ N)$, we say that $(M\ N)$ is an *application* of M to N.

(3) If M is a λ-term and x is a variable, then $(\lambda x \bullet M)$ is also a λ-term. In this circumstance, $(\lambda x \bullet M)$ is said to be an *abstraction*, just because we form a function out of a term (expression).

It is useful in this discussion to follow the convention that application of λ-terms associates to the left. This means that the λ-term $P\,Q\,R\,S$ is to be interpreted as

$$(((P\,Q)R)S).$$

We also treat $(\lambda x \bullet P\,Q)$ as $((\lambda x \bullet P)Q)$.

Following [NIS], we present Table 5.1, which illustrates the most elementary use of the λ formalism. For this first treatment, let us understand a function $\mathit{diff} :: \mathbb{N} \to \mathbb{N} \to \mathbb{N}$ as follows:

$$\mathit{diff}\ x\,y = x - y.$$

Also the function $\mathit{square} :: \mathbb{N} \to \mathbb{N}$ is construed as

$$\mathit{square}\ x = x \times x.$$

Table 5.2 provides some examples of familiar functions and their cognate application in the language of Lambda abstraction. We encourage the reader to work through each example to establish familiarity with the Lambda formalism.

Table 5.3 shows that parentheses are superfluous for a clear understanding of λ-abstraction.

We illustrate (Table 5.4) how function *application* works in the Lambda abstraction.

Again following Nissanke [NIS], we give some examples (Table 5.5) illustrating λ-terms. This display gives a glimpse of the Lambda calculus.

λ-abstraction	**Role of the Variables**
$\lambda\, b \bullet \mathit{diff}(\mathit{square}\ \ a)(\mathit{square}\ \ b)$	a is a free variable and b is a formal parameter.
$\lambda\, a \bullet \mathit{diff}(\mathit{square}\ \ a)(\mathit{square}\ \ b)$	a is a formal parameter and b is a free variable.
$\lambda\, a,\ \ b \bullet \mathit{diff}(\mathit{square}\ \ a)(\mathit{square}\ \ b)$	Both a and b are formal parameters; a is the first and b is the second.
$\lambda\, b,\ \ a \bullet \mathit{diff}(\mathit{square}\ \ b)(\mathit{square}\ \ a)$	Both a and b are formal parameters; b is the first and a is the second.

Table 5.1

Familiar Function	**Lambda Abstraction**
$f(x) = x - 3$ on $\mathbb{N}$	$\lambda\ \ x : \mathbb{N} \bullet x - 3$
$f(x) = \sqrt{x}$ on $\mathbb{Z}$	$\lambda\ \ x : (n \in \mathbb{Z} \wedge n^2 = x) \bullet n$
$f(x, y) = 2x - 3y$ from $\mathbb{R}^2$ to $\mathbb{R}$	$\lambda\, x,\ y : \mathbb{R} \bullet 2x - 3y$
$f(x, y) = \sqrt{x^4 + y^2}$ from $\mathbb{R}^2$ to $\mathbb{R}$	$\lambda\, x,\ y : \mathbb{R} \bullet (x^4 + y^2)^{1/2}$
$f(x, y) = \sqrt{3x + 5y}$ from $\mathbb{R}^2$ to $\mathbb{R}$, if $3x + 5y \geq 0$	$\lambda\, x,\ y : \mathbb{R} \vert 3x + 5y \geq 0 \bullet (3x + 5y)^{1/2}$

Table 5.2

Function Application	**Out**	**Remarks**
$\lambda x \bullet x\, a$	a	$\lambda x \bullet x$ is the identity function.
$\lambda y \bullet x\, a$	x	$\lambda y \bullet x$ is the constant function with value x.
$((\lambda a, b \bullet \mathit{diff}(\mathit{square}\ a)\ (\mathit{square}\ b))5)3$	16	The function is identical to $\lambda x \bullet 16$.

Table 5.3

$\lambda\ \ x : \mathbb{N} \bullet x - 3$	$(\lambda\ \ x : \mathbb{N} \bullet x - 3)5 = 5 - 3 = 2$
$\lambda\ \ x : \mathbb{Z} \mid n \in \mathbb{Z} \wedge n^2 = x \bullet n$	$(\lambda\ \ x : \mathbb{Z} \mid n \in \mathbb{Z} \wedge n^2 = x \bullet n)16 = 4$
$\lambda\ \ x : \mathbb{Z} \mid n \in \mathbb{Z} \wedge n^2 = x \bullet n$	$(\lambda\ \ x : \mathbb{Z} \mid n \in \mathbb{Z} \wedge n^2 = x \bullet n)18 =$ 'undefined'
$\lambda\ \ x, y : \mathbb{R} \bullet 2x - 3y$	$(\lambda\ \ x, y : \mathbb{R} \bullet 2x - 3y)4 = \lambda y : \mathbb{R} \bullet 8 - 3y$
$\lambda\ \ x, y : \mathbb{R} \bullet (x^4 + y^2)^{1/2}$	$(\lambda\ \ x, y : \mathbb{R} \bullet (x^4 + y^2)^{1/2})3 = \lambda y : \mathbb{R} \bullet (81 + y^2)^{1/2}$

Table 5.4

λ-terms	**Meaning**
x	x is a variable (an atom) and hence, according to **(1)** of the syntax, it is a λ-term.
y	y is a variable (an atom) and hence, according to **(1)** of the syntax, it is a λ-term.
$\lambda\ y \bullet y$	y is a λ-term and is also a variable and hence, by **(3)** of the syntax, $(\lambda\ y \bullet y)$ is a λ-term.
$\lambda\ y \bullet x$	x is a λ-term and y is a variable and hence, by **(3)** of the syntax, $(\lambda\ y \bullet x)$ is a λ-term.
$(((\lambda y \bullet y \lambda y \bullet x)y)a)$	This is a λ-term because: $$\underbrace{\underbrace{(((\underbrace{\lambda y \bullet y}_{\substack{\text{abstraction}\\ \text{see}(\mathbf{3})}} \quad \underbrace{\lambda y \bullet x}_{\substack{\text{abstraction}\\ \text{see}(\mathbf{3})}})y)}_{\text{application, see }(\mathbf{2})}a)}_{\text{application, see }(\mathbf{2})}$$
$(\lambda y \bullet x(\lambda y \bullet y a))$	This is a λ-term because: $$\underbrace{(\underbrace{\lambda y \bullet x}_{\text{abstraction}} \; \underbrace{(\underbrace{\lambda y \bullet y}_{\text{abstraction}} \; a))}_{\text{application}}}_{\text{application}}$$ This λ-term represents composition of functions.

Table 5.5

5.11.1 Free and Bound Variables in the λ-Calculus

Although the notion of free and bound variables in the λ-calculus is similar to that in first-order logic (Section 2.3), there are some differences to be noted:

(1) The role of the quantifiers $\forall$ and $\exists$ is played by the symbol λ.

(2) The role of variables and constants in predicate logic is now played by variables and constants in the λ-calculus—that is, by the atoms.

(3) When we determine the scope of a λ-term in the absence of parentheses, especially in the case of terms involving λ, left associativity of application of λ-terms must be figured in.

(4) In pure λ-calculus, there are no analogs to predicate symbols.

Definition 5.7 Let x be a variable in a λ-term M. We say that x is a *bound variable* if it occurs within a part of M having the form $\lambda\, x \bullet N$. Otherwise we say that the occurrence of x is free.

5.11.2 Substitution

Reasoning about λ-terms is effected by their transformation to other λ-terms. This, in turn, is achieved by way of replacement of a variable by a λ-term.

Definition 5.8 The notation

$$E[x/M]$$

denotes the λ-term obtained by replacing all free occurrences of the variable x in the λ-term E by the λ-term M.

We conclude this discussion of the λ-calculus by indicating some rules on substitution of variables:

(1) $y[y/M] \equiv M$.

(2) $x[y/M] \equiv x$ provided that $y \neq x$.

(3) $(E\ F)[y/M] \equiv ((E[y/M])\,(F[y/M]))$.

(4) $(\lambda\, y \bullet E)[y/M] \equiv \lambda\, y \bullet E$.

(5) $(\lambda\, x \bullet E)[y/M] \equiv \lambda x \bullet (E[y/M])$,
provided that:

(a) $y \neq x$;

(b) either x is not free in M or y is not free in E.

(6) $(\lambda x \bullet E)[y/M] \equiv \lambda \bullet (E[x/w][y/M])$,
provided that:

(a) $y \neq x$.

(b) x is free in M.

(c) y is free in E.

(d) w is a "fresh" free variable, in the sense that w is not free in E nor in M.

We refer the reader to [NIS, Chapter 20] for a more detailed treatment of the λ-calculus, together with some applications to computer science. An even more thorough treatment appears in [BAR] and [CF].

5.11.3 Examples

We present finally a couple of examples that exhibit transformations and simplifications of λ-terms. Again, we reference Nissanke [NIS] as our source.

Example 5.26

$$
\begin{aligned}
&(\lambda x \bullet y(\lambda y \bullet y))[y/(\lambda x \bullet y x)] \\
&\equiv ((\lambda x \bullet \underbrace{y}_{E})[y/\underbrace{(\lambda x \bullet y x)}_{M}] \\
&\quad (\lambda y \bullet y)[y/(\lambda x \bullet y x)]) && \textbf{rule(3)} \\
&\equiv (\lambda p \bullet y\,[x/p][y/(\lambda x \bullet y x)] \\
&\quad (\lambda y \bullet y)[y/\lambda x \bullet y x)]) && \textbf{rule(6)} \\
&\equiv (\lambda p \bullet y[y/(\lambda x \bullet y x)] \\
&\quad (\lambda y \bullet y)[y/(\lambda x \bullet y x)]) && \textbf{rule(2)} \\
&\equiv (\lambda p \bullet (\lambda x \bullet y x)(\lambda y \bullet y) \\
&\quad [y/(\lambda x \bullet y x)]) && \textbf{rule(1)} \\
&\equiv (\lambda p \bullet (\lambda x \bullet y x)(\lambda y \bullet y)) && \textbf{rule(4)}
\end{aligned}
$$

□

Example 5.27

$$
\begin{array}{ll}
(\lambda x \bullet (y(\lambda y \bullet y)))[y/(\lambda x \bullet (y\,x))] & \\
\quad \equiv (\lambda x \bullet (y(\lambda y \bullet y)) & \\
\qquad [y/(\lambda x \bullet (y\,x))]) & \textbf{rule}(\textbf{5}) \\
\quad \equiv (\lambda x \bullet (y[y/(\lambda x \bullet (y\,x))] & \\
\qquad (\lambda y \bullet y)[y/(\lambda x \bullet (y\,x))])) & \textbf{rule}(\textbf{3}) \\
\quad \equiv (\lambda x \bullet ((\lambda x \bullet (y\,x))(\lambda y \bullet \underbrace{y}_{E}) & \\
\qquad [y/\underbrace{(\lambda x \bullet (y\,x))}_{M}])) & \textbf{rule}(\textbf{1}) \\
\quad \equiv (\lambda x \bullet ((\lambda x \bullet (y\,x))(\lambda y \bullet y))) & \textbf{rule}(\textbf{4})
\end{array}
$$

□

5.12 Sequences

A *sequence* on a set S is a function $\varphi : \mathbb{N} \to S$. We often write $s_1, s_2, \ldots$ or $\{s_j\}_{j=1}^{\infty}$. For example,

$$2, 4, 8, \ldots, 2^j, \ldots$$

is a sequence on the integers.

In computer science, it is often convenient to use finite sequences. A finite sequence is a partial function from $\mathbb{N}$ to S (i.e., from some set $\{1, 2, \ldots, N\}$ to S). For instance,

$$3, 7, 10, 35, 92$$

is a finite sequence on the integers. Of course there is no requirement that a sequence conform to some pattern or rule, but often it will.

5.13 Bags

A *bag* is a set in which multiple occurrences of objects are allowed. A bag is also called a multi-set. A typical example of a bag is

$$d = \{(\text{book}, 4), (\text{rat}, 6), (\text{plum}, 2), (\text{gun}, 5)\}.$$

This notation specifies a bag that contains 4 books, 6 rats, 2 plums, and 5 guns. We also write

$$d = [[\text{book}, \text{rat}, \text{plum}, \text{gun}]]$$

to enumerate the elements of the bag without specifying the number of instances of each.

There are several standard operations on bags:

- "count" for specifying the number of some element;
- # for counting;
- $\uplus$ for specifying "bag-union";
- $\bigcup\!\!\!-$ for specifying "bag-difference";
- E for specifying "element of";
- $\sqsubseteq$ for specifying "subset of."

We now illustrate these with simple examples. Let

$$c = \{(\text{book}, 3), (\text{rat}, 7), (\text{plum}, 2), (\text{cow}, 3), (\text{horn}, 7)\},$$

$$d = \{(\text{book}, 4), (\text{rat}, 4), (\text{plum}, 2), (\text{gun}, 5)\},$$

$$e = \{(\text{book}, 3), (\text{rat}, 5), (\text{gun}, 5)\},$$

$$f = \{(\text{book}, 3), (\text{rat}, 8), (\text{plum}, 3), (\text{cow}, 3), (\text{horn}, 9), (\text{lamp}, 2)\}.$$

Then

$$\text{count } d \text{ rat} = 4 \qquad \text{count } f \text{ plum} = 3$$

because there are 4 rats in the bag d and 3 plums in the bag f.

Also

$$e\#\text{gun} = \text{count } e \text{ gun} = 5 \qquad c\#\text{cow} = \text{count } c \text{ cow} = 3$$

because there are 5 guns in the bag e and 3 cows in the bag c.

Next,

$$d \uplus e = \{(\text{book}, 4), (\text{rat}, 5), (\text{gun}, 5), (\text{plum}, 2)\},$$

$$c \,\bigcup\!\!\!-\, d = \{(\text{rat}, 3), (\text{cow}, 3), (\text{horn}, 7)\}.$$

Observe that numbers of individual elements subtract in an obvious way, just as in the set-theoretic difference.

We write

$$\text{rat} \,\mathsf{E}\, c \quad \text{and} \quad \text{horn} \,\not{\mathsf{E}}\, d$$

to denote "element of" and "not an element of."

Finally, we write

$$c \sqsubseteq f \quad \text{and } c \not\sqsubseteq d$$

to indicate that c is a bag-subset of f in an obvious way (i.e., for any given element, f has at least as many as c) and, likewise, that c is not a bag-subset of d.

Chapter 6

Recursive Functions

We need education in the obvious more than investigation of the obscure.

—O.W. Holmes, II

To reverence superiority and accept a fact though it slay him are the final tests of an educated man.

—Martin H. Fischer

Mathematics has not a foot to stand upon which is not purely metaphysical. It begins in metaphysics; and their several orbits are continually intersecting.

—Thomas de Quincey

A small inaccuracy can save hours of explanation.

—H.H. Munro (Saki)

The validity of mathematical propositions is independent of the actual world—the world of existing subject-matters—, is logically prior to it, and would remain unaffected were it to vanish from being.

—Cassius J. Keyser

Philosophy simply puts everything before us and neither explains nor deduces anything. Since everything lies open to view, there is nothing to explain. For what is hidden, for example, is of no interest to us.

—Ludwig Wittgenstein

What men really want is not knowledge but certainty.

—Bertrand Russell

Logic is a gamble, at terrible odds—if it was a bet you wouldn't take it.

—Tom Stoppard

6.1 Introductory Remarks

The point at issue in the study of recursive functions is to determine which functions are effectively computable. Church's thesis asserts that the collection of effectively computable functions is precisely the collection of general recursive functions. One knows that a function is effectively computable only after one has effectively computed it. By contrast, the general recursive functions are generated from well-defined beginnings using very strict rules of construction, so the issue that is begged is one of explicit construction versus abstract existence.

Here we give some detailed definitions and examples and discuss the problem at hand. We of course provide no proofs, but refer the reader to [COH], [END], and [SCH] for a more thoroughgoing discussion.

6.1.1 A System for Number Theory

We will carry out our discussion in the context of an axiom system that P. J. Cohen calls Z_2. It is a system for basic number theory that disallows the formation of infinite sets:

1. $\forall x, \forall y, [x = y] \iff \forall z, [z \in x \iff z \in y]$;
2. $\forall x, [\sim x \in \emptyset]$;
3. $\forall x, \forall y, \exists z, \forall w, (w \in z \iff [w = x \vee w = y])$;
4. $\forall x, \forall y, \exists z, \forall w, [w \in z \iff w \in x \vee w \in y]$.

We then define a number x to be an integer if

1. $\forall y, \forall z, [y \in x \wedge z \in x] \Rightarrow [y = z \vee y \in z \vee z \in y]$;
2. $\forall y, \forall z, [y \in x \wedge z \in y] \Rightarrow z \in x$.

Finally, if x is an integer, then we let $x + 1$ denote $x \cup \{x\}$ (as in the theory of ordinal numbers—see Subsection 5.8.10).

6.2 Primitive Recursive Functions

A function $f(n_1, \ldots, n_k)$ from $\mathbb{Z}$ to $\mathbb{Z}$ is called *primitive recursive* (p.r.) if it is constructed by means of the following rules.

1. $f \equiv c$ for some constant c is p.r.
2. $f(n_1, \ldots, n_k) = n_i$, for some $1 \leq i \leq k$, is p.r.
3. $f(n) = n + 1$ is p.r.
4. If $f(n_1, \ldots, n_k)$ and $g_1, \ldots, g_k$ are p.r., then so is $f(g_1, \ldots, g_k)$.
5. If $f(0, n_2, \ldots, n_)$ is p.r. and if $g(m, n_1, \ldots, n_k)$ is p.r. and if we have $f(n_1, n_2, \ldots, n_k) = g(f(n, n_2, \ldots, n_k), n, n_2, \ldots, n_k)$, then f is p.r.

The elementary functions of the integers that one typically encounters—addition, multiplication, powers, factorials, the nth prime—are certainly primitive recursive.

6.2.1 Effective Computability

The important thing about the p.r. functions is that they are *effectively computable*. Here a function or procedure **m** is said to be effectively computable if

1. The procedure **m** is finite in length and time.
2. The procedure **m** is fully explicit and nonambiguous.
3. The procedure **m** is faultless and infallible.
4. The procedure **m** can be carried out by a machine.

A careful discussion of all of these conditions is carried out by Luciano Floridi at `http://www.wolfson.ox.ac.uk/~floridi/ctt.htm`. For the fourth condition, the meaning is that the steps to construct the function should consist of mindless adherence to certain explicit rules that can be encoded so that they can be understood by a computing machine. The rules that we have given for creating p.r. functions clearly are of that nature.

6.2.2 Effectively Computable Functions and p.r. Functions

In view of Church's thesis, it is important to note that the p.r. functions do not exhaust the class of effectively computable functions. To see this, let $f_m(n)$ be a list of all schemes for generating p.r. functions of one variable. This is an effectively computable function of the two variables m and n. Now set $g(n) = f_n(n) + 1$. Then g is *not* a p.r. function, but it is effectively computable (because we have just effectively computed it!), so the p.r. functions are the simplest class of computable functions; they are just not all of the (effectively) computable functions.

An important result that clarifies the role of p.r. functions is the following.

Theorem 6.1
If f is a p.r. function, then there is a formula A in Z_1 (see Subsection 3.4.3, and contrast with the definition of Z_2) such that $f(n_1, \ldots, n_k) = x$ if and only if $A(n_1, \ldots, n_k, x)$.

6.3 General Recursive Functions

Recursive functions were first defined by Gödel. To define these functions, we must first introduce our "alphabet." It will consist of:

- function symbols $f, g, h, \ldots$, each associated with a fixed finite number of variables;

- integer variables $x, y, z, \ldots$;
- the symbols 0 (zero) and $'$ (the successor: x' represents $x+1$);
- parentheses (,) and the comma , .

Now we need some primitive definitions:

1. A *numeral* is an expression of the form $0, 0', 0,'' \ldots$;
2. A *term* is defined as follows.
   (a) The symbol 0 denotes a term.
   (b) The variables $x, y, z, \ldots$ denote terms.
   (c) If r is a term, then so is r'.
   (d) If $r_1, \ldots, r_n$ are terms and if f is a function symbol of n variables, then $f(r_1, \ldots, r_n)$ is a term.
3. An *equation* is an expression of the form $r = s$, where r and s are terms.
4. If E is a finite set of equations, then an equation is a *deduction* from E if it can be obtained by repeated applications of the following rules:
   (a) Given an equation A, we may replace all occurrences of a given variable x by a given numeral.
   (b) If $f(n_1, \ldots, n_k) = m$ has been deduced, where m and n_i are numerals, then given any equation we may replace an occurrence of $f(n_1, \ldots, n_k)$ in that equation by m.
   (c) If $r = s$ is deduced, then $s = r$ is deduced.

Definition 6.1 A function $f(x_1, \ldots, x_k)$ is said to be *general recursive* (or, sometimes, just "recursive") if there is a finite set of equations (expressed in terms of f) such that, for any choice of the numerals $n_1, \ldots, n_k$, there is a unique m such that $f(n_1, \ldots, n_k) = m$ can be deduced.

Two simple examples follow.

Example 6.1

Using, as usual, $'$ to denote the successor, we consider the set of equations

$$f(x, 0) = x \qquad f(x, y') = f(x, y)'.$$

This defines the addition operation, $f = x + y$, and shows that addition is a recursive function. (More generally, this paradigm can be used to show that any primitive recursive function is general recursive.)

□

Example 6.2

Again, we use $'$ to denote the successor operation. Set

$$f(0, m') = m',$$
$$f(n, 0) = n',$$
$$f(n', m') = f(n, f(n', m)).$$

Thus f is a recursive function. Moreover, if g is any primitive recursive function, then, for some x, $f(x, y) > g(y)$ for every y (showing that f is not primitive recursive).

□

6.3.1 Every Primitive Recursive Function Is General Recursive

Certainly every primitive recursive function is general recursive, as the reader may easily verify. An example of a p.r. function cast in recursive language is to set $f(x, 0) = x$ and $f(x, y') = [f(x, y)]'$. (Here, as usual, the $'$ denotes the successor operation.) This defines $f \equiv x + y$ (see also Example 6.1).

We reiterate: Church's thesis is that the class of general recursive functions exhausts all of the effectively computable functions. This is a statement of philosophy; it is not amenable to proof. The literature on Church's thesis consists largely of rather free-ranging discussions of the meaning of the concept of "effective computability."

6.3.2 Turing Machines

A *Turing machine* is a device for performing effectively computable operations. It consists of a machine through which a bi-infinite paper tape is fed. The tape is divided into an infinite sequence of congruent boxes (Figure 6.1). Each box has either a numeral 0 or a numeral 1 in it. The Turing machine has finitely many "states" $S_1, S_2, \ldots, S_n$. In any given state of the Turing machine, one of the boxes is being scanned.

After scanning the designated box, the Turing machine does one of three things:

(1) It either erases the numeral 1 that appears in the scanned box and replaces it with a 0, or it erases the numeral 0 that appears

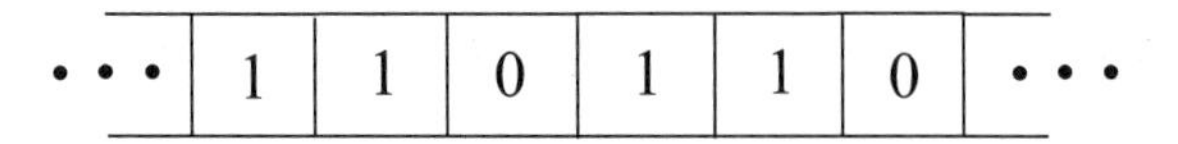

Figure 6.1

in the scanned box and replaces it with a 1, or it leaves the box unchanged.

(2) It moves the tape one box (or one unit) to the left or to the right.

(3) It goes from its current state S_j into a new state S_k.

6.3.3 An Example of a Turing Machine

Here is an example of a Turing machine for calculating $x + y$:

State	Old Value	New Value	Move (l. or r.)	New State	Explanation
0	1	1	R	0	pass over x
0	0	1	R	1	fill gap
1	1	1	R	1	pass over y
1	0	0	L	2	end of y
2	1	0	L	3	erase a 1
3	1	0	L	4	erase another 1
4	1	1	L	4	back up
4	0	0	R	5	halt

6.3.4 Turing Machines and Recursive Functions

Now let us think about the Turing machine in the language of recursive functions.

To describe stipulation (1), we need a function $\varphi_1(i, j)$, where $i = 0, 1$ and $j = 1, 2, \ldots, n$. If the machine is in state S_j and if the symbol 0 appears in the scanned box, then $\varphi(0, j) = 0$ if the box is left unchanged and $\varphi(0, j) = 1$ if the 0 is erased and a 1 is written in the box. If the machine is in state S_j and if the numeral 1 appears in the scanned box, then $\varphi(1, j) = 0$ if the box is left unchanged and $\varphi(0, j) = 1$ if the numeral 1 is erased and replaced with a 0.

To describe stipulation (2), we need a function $\varphi_2(i, j)$, where $i = 0, 1$ and $j = 1, 2, \ldots, n$. If the machine is in state S_j and if the symbol 0 appears in the scanned box, then $\varphi_2(0, j) = 0$ if the tape shifts to the left and $\varphi_2(0, j) = 1$ if the tape shifts to the right. If the machine is in state S_j and if the numeral 1 appears in the scanned box, then $\varphi_2(1, j) = 0$ if the tape shifts to the left and $\varphi_2(1, j) = 1$ if the tape shifts to the right.

To describe stipulation (3), we need a function $\varphi_3(i, j)$, where again $i = 0, 1$ and $j = 1, 2, \ldots, n$. The function φ_3 will take the values $1, 2, \ldots, n$. If the scanned box has a zero and the machine is in state S_j, then the value of $\varphi_3(0, j)$ is the index of the new state that the machine will enter. If the scanned box contains the numeral 1 and the machine is in state S_j, then the value of $\varphi_3(1, j)$ is the index of the new state that the machine will enter.

One can imagine a "super" Turing machine M that will be given two numbers as input, one of them a suitable encoding of a standard Turing machine N and the other a number x. Machine M might then act as an "umbrella" or "executive program"; the output might then be just the result of applying standard Turing machine N to the data x. We can then call M a *universal Turing machine.*

6.3.5 Defining a Function with a Turing Machine

Now we describe how one uses the Turing machine to define a function. Let m be an integer. We will define $f(m)$ for a function f. Imagine beginning with a tape in which m consecutive boxes are occupied with the numeral 1. We start the Turing machine so that it scans the leftmost of these occupied boxes, and so that it is in the initial state S_1. Now we run the Turing machine according to the three rules given above. If the machine reaches a state S_k that it never leaves, and in which it never alters the tape, then we say that the calculation has ended—*provided* that, in addition, the tape has all zeros except for a consecutive string of boxes with 1's and the leftmost of those boxes with 1's is the one being scanned. The number of boxes with 1's in that terminal state is declared to be the value $f(m)$.

It is an important theorem (see [KLE]) that the class of functions that can be calculated by Turing machines is just the same as the class of general recursive functions.

6.3.6 Recursive Sets

If S is any set of integers, we let the *characteristic function* of S be

$$\chi_S(x) = \begin{cases} 1 \text{ if } x \in S \\ 0 \text{ if } x \notin S. \end{cases}$$

Definition 6.2 A set S is said to be *recursive* if its characteristic function χ_S is general recursive.

6.3.7 Recursively Enumerable Sets

Definition 6.3 A set S is *recursively enumerable* if S is either empty or is the image of a general recursive function.

Example 6.3

For a fixed k, enumerate all pairs consisting of a polynomial $p(x_1, \dots, x_k)$ with integral coefficients together with a k-tuple $\mathbf{n} = (n_1, \dots, n_k)$ of integers. If $(p, \mathbf{n})$ is an element of this sequence (i.e., an ordered pair of a polynomial and a k-tuple of integers), define $f(p, \mathbf{n})$ to be equal to $p(x_1, \dots, x_k)$ if $p(n_1, \dots, n_k) = 0$ and otherwise let $f(p, \mathbf{n}) = 0$. Then the image $\mathbf{P}$ of this function f is the set of all polynomials $p(x_1, \dots, x_k)$ with integer coefficients that have an integer solution. It is not known whether $\mathbf{P}$ is recursively enumerable, that is, we do not know whether f, or some function like f, is general recursive.

□

The principal result of the subject of general recursive functions is formulated in the following result.

Theorem 6.2
There exists a recursively enumerable set that is not recursive.

6.3.8 The Decision Problem

The *decision problem* is to determine whether, given any statement A in Z_2, there is an effective method (discussed below) that yields a proof of either A or $\sim A$. As a corollary of the last theorem, it can be proved that the decision problem is unsolvable. It can further be shown that Diophantine equations (polynomial equations with integer coefficients for which we seek integer solutions) have an unsolvable decision problem (treated below).

More generally, the decision problem (see [MEN, pp. 254–255]) is this: Given a class of problems, is there a general effective procedure for solving each problem in the given class? If it is possible to arithmetize the formulations of a general class of problems and therefore assign to each problem a natural number, then the class is undecidable if and only if there is no effectively computable function k such that, if n is the number or index of a given problem, then $k(n)$ gives a solution of the problem. Accepting Church's thesis (Subsection 3.5.2, Section 6.1), the function k would have to be general recursive in order to obtain effective computability.

6.3.9 Decision Problems with Negative Resolution

Examples of decision problems that have been resolved negatively are

- the word problem for semi-groups [POS], [KLE];
- the word problem for groups [BOO], [NOV], [BRIT], [HIG] (Section 14.4);
- the satisfiability problem [COO] (Subsection 12.8.1);
- the problem of recognizing which Diophantine equations have solutions [DAV2] (Subsection 6.3.7).

The *method of problem reduction* consists simply of reducing a problem to another problem that is known to be undecidable. One of the most frequently occurring problems that is known to be undecidable is the Post correspondence problem:

> **[Post]** Consider a finite sequence of ordered pairs (s_1, t_1), (s_2, t_2), $\ldots$, (s_k, t_k). Here the s_j, t_j are binary strings of positive length. Is there a sequence of indices $i_1, i_2, \ldots, i_m$ with the property that the concatenation of strings given by $s_{i_1} s_{i_2} \cdots s_{i_m}$ equals the concatenation of strings given by $t_{i_1} t_{i_2} \cdots t_{i_m}$?

As an example, the question of the decidability of satisfaction in predicate logic can be reduced to Post's problem. This result was established by Church.

The following remarkable result simplifies some calculations.

Theorem 6.3
Every nonempty recursively enumerable set is actually the image of a primitive recursive function.

By way of addressing Church's thesis (but certainly not establishing it as a proven fact), we have the next result about the μ-operator.

6.3.10 The μ-Operator

The μ-operator is defined as follows.

Definition 6.4 If $f(y, x_1, \ldots, x_n)$ is any function, then let

$$\mu f(y, x_1, \ldots, x_n) = \mu_y f(y, x_1, \ldots, x_n)$$

denote the function $g(x_1, \ldots, x_n)$ defined by these rules:

- We set $g(x_1, \ldots, x_n) = 0$ if, for any $\overline{x}_1, \ldots, \overline{x}_n$ (where $\overline{x}_1, \ldots \overline{x}_n$ are the images of $x_1, \ldots, x_n$ in some model), we have the nonvanishing of f: $f(y, \overline{x}_1, \ldots, \overline{x}_n) \neq 0$ for every y.

- If $x_1, \ldots, x_n$ are entries such that $f(y, \overline{x}_1, \ldots, \overline{x}_n) = 0$ for some y, then we set $g(x_1, \ldots, x_n) = a$, where a is the *least* entry y with $f(y, \overline{x}_1, \ldots, \overline{x}_n) = 0$.

Theorem 6.4
The general recursive functions form the smallest class of functions that contains the primitive recursive functions and is closed under composition and the μ-operator.

Chapter 7

The Number Systems

... there are good reasons to believe that nonstandard analysis, in some version or other, will be the analysis of the future.

—Kurt Gödel

Sire, I have no need of that hypothesis.

—Pierre Simon de Laplace

I must create a system or be enslaved by another man's.

—William Blake

We ought not to be over-anxious to encourage innovation in cases of doubtful improvement, for an old system must ever have two advantages over a new one: it is established, and it is understood.

—Charles Caleb Colton

I frame no hypothesis; for whatever is not deduced from the phenomena is to be called an hypothesis; and hypotheses, whether metaphysical or physical, whether of occult qualities or mechanical, have no place in experimental philosophy.

—Isaac Newton

Logical consequences are the scarecrows of fools and the beacons of wise men.

—T.H. Huxley

Errors are not in the art but in the artificers.

—Isaac Newton

Man is a credulous animal and must believe something. In the absence of good grounds for belief, he will be satisfied with bad ones.

—Bertrand Russell

Logic is simply the architecture of human reason.

—Evelyn Waugh

7.1 The Natural Numbers

7.1.1 Introductory Remarks

It is in fact quite difficult to construct—from first principles—a system of natural numbers in which the multiplication operation is workable.

In many treatments, an extra axiom is added in order to make multiplication have the properties that we want it to have (see [SUP2, p. 136]). In other treatments, the natural numbers are taken as undefinables. We will take a third approach, which adheres more closely to the philosophy of ordinal numbers.

7.1.2 Construction of the Natural Numbers

We inductively construct numbers as follows

$$\begin{aligned} \mathbf{0} &= \emptyset \\ \mathbf{1} &= \{\emptyset\} \\ \mathbf{2} &= \{\emptyset, \{\emptyset\}\} \\ &\cdots \\ \mathbf{n+1} &= n \cup \{n\} \\ &\cdots. \end{aligned}$$

These are the natural numbers. The set of natural numbers is denoted by $\mathbb{N}$. We commonly enumerate them as $0, 1, 2, \ldots$.

What is important about the natural numbers, indeed about any number system, is its closure under certain arithmetic operations. Our particular construction of the natural numbers lends itself well to verifying this property for addition. We define addition inductively:

$$\begin{aligned} n+1 &= n \cup \{n\}; \\ n+2 &= (n+1)+1 \\ &\text{etc.} \end{aligned}$$

For example,

$$\begin{aligned} 2+2 = (2+1)+1 &= \Big[\{\emptyset, \{\emptyset\}\} \cup \Big\{\{\emptyset, \{\emptyset\}\}\Big\}\Big] + 1 \\ &= \Big\{\emptyset, \{\emptyset\}, \{\emptyset, \{\emptyset\}\}\Big\} \cup \Bigg\{\Big\{\emptyset, \{\emptyset\}, \{\emptyset, \{\emptyset\}\}\Big\}\Bigg\} \\ &= \Big\{\emptyset, \{\emptyset\}, \{\emptyset, \{\emptyset\}\}, \{\emptyset, \{\emptyset\}, \{\emptyset, \{\emptyset\}\}\}\Big\} \\ &= 4. \end{aligned}$$

It is convenient in the logical construction of the natural numbers to include zero as a natural number. In particular, our definition makes the additive law

$$n + 0 = n \cup 0 = n \cup \emptyset = n$$

very natural. But the reader should be warned that, in common mathematical parlance, the name "natural numbers" and the symbol $\mathbb{N}$ are generally reserved for the set $\{1, 2, 3, \ldots\}$ of *positive* integers. In common mathematical discourse, the set $\{0, 1, 2, \ldots\}$ is generally denoted by $\mathbb{Z}^+$ and is called "the nonnegative integers."

7.1.3 Axiomatic Treatment of the Natural Numbers

In practice, it is most convenient to treat the natural numbers axiomatically. Guiseppe Peano (1858–1932) formulated the axiomatic theory of the natural numbers that we still use today. His axioms are these:

1. Each natural number has a unique successor.
2. There is a natural number 1 that is not the successor of any natural number.
3. Two distinct integers cannot have the same successor.
4. If M is a set of natural numbers such that $1 \in M$ and such that if a natural number n is in M then its successor is also in M, then every natural number is in M.

Observe that the first axiom guarantees that there are infinitely many natural numbers, and they are the the ones we expect. Axiom 2 guarantees that there is a first natural number. Axiom 3 requires no comment. Axiom 4 amounts to the principle of mathematical induction.

Peano's axiomatic system is simple and complete. But it does *not* provide the machinery for addition or multiplication. In fact it is an as yet unresolved problem to determine definitively how to perform the usual arithmetic operations in Peano's system. The customary method for handling addition and multiplication is to add some axioms. We will not explore the details here, but refer the reader to [SUP2, p. 136].

7.2 The Integers

7.2.1 Lack of Closure in the Natural Numbers

The natural numbers are closed under addition and multiplication. They are *not* closed under subtraction (for example, $3 - 5$ is not an element of the natural numbers). To achieve closure under that new operation, we must expand the number system as follows.

Let

$$X = \{(m, n) : m, n \in \mathbb{N}\} \equiv \mathbb{N} \times \mathbb{N}.$$

We define a relation on X by

$$(m, n) \sim (m', n') \qquad \text{if and only if} \qquad m + n' = m' + n.$$

It is straightforward to verify that this is an equivalence relation, and we leave the details to the reader.

7.2.2 The Integers as a Set of Equivalence Classes

Now the set of equivalence classes of X (we denote the set of equivalence classes by $X/\sim$), under this equivalence relation, is the new number system that we will call the *integers* (denoted by $\mathbb{Z}$, from the German word *Zahl* for number). In fact, we think of the integer that we commonly denote by k (for $k \geq 0$) as the equivalence class $\{(m,n) : m \in \mathbb{N}, n \in \mathbb{N}, m+k=n\}$. For $k<0$ (assuming that the reader is familiar with the ordinary arithmetic of negative integers), we think of k as the equivalence class $\{(m,n) : m \in \mathbb{N}, n \in \mathbb{N}, m=n-k\}$. Our rules of arithmetic in this new number system are

- $[(m,n)] + [(k,\ell)] = [(m+k, n+\ell)]$;
- $[(m,n)] - [(k,\ell)] = [(m+\ell, n+k)]$;
- $[(m,n)] \cdot [(k,\ell)] = [(m\ell+nk, n\ell+mk)]$.

7.2.3 Examples of Integer Arithmetic

These definitions are best understood by way of some examples:

$$\begin{aligned} 3+(-5) &= [(1,4)] + [(9,4)] = [(1+9, 4+4)] = [(10,8)] = -2 \\ 4-8 &= [(2,6)] - [(1,9)] = [(2+9, 6+1)] = [(11,7)] = -4 \\ 3\cdot(-6) &= [(2,5)] \cdot [(10,4)] = [(2\cdot 4 + 5\cdot 10, 2\cdot 10 + 5\cdot 4)] \\ &= [(58,40)] = -18. \end{aligned}$$

7.2.4 Arithmetic Properties of the Negative Numbers

The satisfying thing about the construction given here is that the standard arithmetic properties of negative numbers are automatic—they are built into the way we have defined our new number system. Observe that the additive identity is $0 = [(1,1)]$ and, indeed, $n+0=n$ for any integer n. Also the multiplicative identity is $1 = [(1,2)]$, and one may check that $1 \cdot n = n$ for any n.

7.3 The Rational Numbers

7.3.1 Lack of Closure in the Integers

The system $\mathbb{Z}$ of integers is closed under addition, subtraction, and multiplication; but it is *not* closed under division. For example, $5 \div 7 = 5/7$

makes no sense in the integers. In order to achieve the desired closure property, we enlarge the system of integers as follows:

Let

$$Y = \{(p, q) : p, q \in \mathbb{Z}, q \neq 0\}.$$

We define a relation on Y by

$$(p, q) \sim (p', q') \qquad \text{if and only if} \qquad p \cdot q' = p' \cdot q.$$

It is straightforward to verify that this is an equivalence relation, and we leave the details to the reader (or see Subsection 5.5.4).

7.3.2 The Rational Numbers as a Set of Equivalence Classes

Now the set of equivalence classes of Y (we denote this set by $Y/\sim$), under this equivalence relation, is the new number system that we will call the *rational numbers* (denoted by $\mathbb{Q}$, where $\mathbb{Q}$ should be considered to be an abbreviation for "quotient"). In fact, we think of the rational that we commonly denote by p/q (in lowest terms) as the equivalence class $\{(pk, qk) : k \in \mathbb{Z}, k \neq 0\}$. Our rules of arithmetic in this new number system are:

- $[(p, q)] + [(r, s)] = [(ps + qr, qs)]$;
- $[(p, q)] \cdot [(r, s)] = [(pr, qs)]$.

7.3.3 Examples of Rational Arithmetic

These definitions are best understood by way of some examples.

$$\begin{aligned}
\frac{3}{5} + \frac{2}{7} &= [(3, 5)] + [(2, 7)] = [(3 \cdot 7 + 5 \cdot 2, 5 \cdot 7)] \\
&= [(31, 35)] = \frac{31}{35} \\
\frac{-4}{9} + \frac{2}{5} &= [(-4, 9)] + [(2, 5)] = [((-4) \cdot 5 + 9 \cdot 2, 9 \cdot 5)] \\
&= [(-2, 45)] = \frac{-2}{45} \\
\frac{-3}{11} \cdot \frac{5}{13} &= [(-3, 11)] \cdot [(5, 13)] = [(-15, 143)] \\
&= \frac{-15}{143}.
\end{aligned}$$

7.3.4 Subtraction and Division of Rational Numbers

The operations of subtraction and division on the rationals are already implicit in addition and multiplication. For completeness, however, we record them here:

- $[(p, q)] - [(r, s)] = [(ps - qr, qs)]$;
- $[(p, q)] \div [(r, s)] = [(ps, qr)]$, provided $r \neq 0$.

7.4 The Real Numbers

7.4.1 Lack of Closure in the Rational Numbers

The set $\mathbb{Q}$ of rational numbers is closed under the standard arithmetic operations of $+, -, \times, \div$. In fact, $\mathbb{Q}$ is a field (see the field axioms in Subsection 7.4.2). From a strictly algebraic point of view, this number system is completely satisfactory for elementary purposes, and in fact it is the rational numbers that are used in everyday commerce, engineering, and science.

However, from a more advanced point of view the rational numbers are not completely satisfactory. This assertion was first discovered by the Pythagoreans more than 2000 years ago (Subsection 8.3.1). Indeed, they determined that there is no rational number whose square is 2 (see Example 8.8). More generally, a positive integer has a rational square root if and only if it has an integer square root. Thus we find that a broader class of numbers is more desirable.

The modern point of view is that the rational numbers are deficient from the point of view of metric topology. More precisely, the sequence of rational numbers given (for instance) by

$$3, 3.1, 3.14, 3.141, 3.1415, 3.14159, \ldots$$

gives better and better approximations to the ratio of the circumference of a circle to its diameter. These numbers are becoming and staying closer and closer together, and appear to converge to some value. But, as it turns out, that value cannot be rational. In fact, the value is π, and it is known that π is not rational. By the same token, the numbers

$$1, 1.4, 1.41, 1.414, 1.4142, \ldots$$

give better and better approximations to the square root of 2, and that number is known to be irrational (Subsection 8.3.1). In summary, we require a system of numbers that is still closed under the basic arithmetic operations, but is also closed under the limiting processes just described.

7.4.2 Axiomatic Treatment of the Real Numbers

In fact, the system of *real numbers* will fill the need just described. It is rather complicated to give a formal construction of the real numbers $\mathbb{R}$, and we refer the reader to [KRA], [RUD], and [STR] for details. We content ourselves here with enunciating an axiom system for the reals (these are taken from [STR]). We state once and for all that it is possible to present an explicit model for a number system that satisfies these axioms (constructed, for example, by the method of Dedekind cuts—see [KRA]). We will take that model for granted, and we will use the real numbers in the remainder of this book.

The real numbers are a field of numbers equipped with a topology that makes the field operations (addition + and multiplication $\cdot$) continuous. Under this topology, the real numbers are *complete*. The detailed axioms are these:

Axiom 1 (Commutative Laws) For all $x, y \in \mathbb{R}$,

$$x + y = y + x \qquad \text{and} \qquad x \cdot y = y \cdot x.$$

Axiom 2 (Associative Laws) For all $x, y, z \in \mathbb{R}$,

$$x + (y + z) = (x + y) + z \qquad \text{and} \qquad x \cdot (y \cdot z) = (x \cdot y) \cdot z.$$

Axiom 3 (Distributive Law) For all $x, y, z \in \mathbb{R}$,

$$x \cdot (y + z) = x \cdot y + x \cdot z.$$

Axiom 4 (Identity Elements) There exist two distinct elements 0 and 1 in $\mathbb{R}$ such that, for all $x \in \mathbb{R}$,

$$0 + x = x \qquad \text{and} \qquad 1 \cdot x = x.$$

Axiom 5 (Inverse Elements) If $x \in \mathbb{R}$, then there exists a unique $-x \in \mathbb{R}$ such that

$$x + (-x) = 0.$$

If $x \in \mathbb{R}$ and $x \neq 0$, then there is a unique element $x^{-1} \in \mathbb{R}$ such that

$$x \cdot x^{-1} = 1.$$

Axiom 6 (Positive Numbers The real number system $\mathbb{R}$ has a distinguished subset $\mathbf{P}$ (the positive numbers) that induces an ordering on $\mathbb{R}$. The three sets $\mathbf{P}$, $\{0\}$, and $-P = \{-x : x \in P\}$ are pairwise disjoint and their union is all of $\mathbb{R}$. We write $a < b$ in case $b - a \in P$.

Axiom 7 (Closure Properties of P) If $x, y \in P$, then $x+y \in P$ and $x \cdot y \in P$.

Axiom 8 (Dedekind Completeness) Let A and B be subsets of $\mathbb{R}$ such that

(i) $A \neq \emptyset$ and $B \neq \emptyset$;

(ii) $A \cup B = \mathbb{R}$;

(iii) $a \in A$ and $b \in B$ imply that $a < b$.

Then there exists exactly one element $x \in \mathbb{R}$ such that

(iv) If $u \in \mathbb{R}$ and $u < x$, then $u \in A$;

(v) If $v \in \mathbb{R}$ and $x < v$, then $v \in B$.

Plainly, the number x described in Axiom 8 must be either in A or in B but not in both. Thus $B = \mathbb{R} \setminus A$ and either $A = \{u \in \mathbb{R} : u \leq x\}$ or $A = \{u \in \mathbb{R} : u < x\}$.

The first seven axioms of the real numbers are also satisfied by the rational numbers. It is these first seven axioms that constitute the postulates for a field. (In some treatments, Axioms 1, 2, 4, and 5 are each split into two; so it is common to say that there are eleven field axioms.)

It is Axiom 8 that makes the real numbers special. It says, in effect, that the real numbers have no gaps or holes in them. In other word, the reals are *complete*. (This use of the word is distinct from the use of "complete" in formal logic—see [BAR, p. 16]. See also Gödel's completeness theorem in Section 1.4.)

7.5 The Complex Numbers

7.5.1 Intuitive View of the Complex Numbers

Intuitive treatments of the complex numbers are unsatisfactory because they posit (without substantiation) the existence of a number that plays the role of the square root of -1. With the formalism developed thus far in this book, we can actually construct the complex numbers.

7.5.2 Definition of the Complex Numbers

We let

$$\mathbb{C} = \{(x, y) : x \in \mathbb{R}, y \in \mathbb{R}\}.$$

We equip $\mathbb{C}$ with the following operations:

$$(x, y) + (x', y') = (x + x', y + y')$$
$$(x, y) \cdot (x', y') = (xx' - yy', xy' + x'y).$$

Notice that $\mathbb{C}$ is not a set of equivalence classes; it is merely a set of ordered pairs. The rule for multiplication may seem artificial, but it is the rule that is needed to turn $\mathbb{C}$ into a field (see Subsection 7.4.2).

7.5.3 The Distinguished Complex Numbers 1 and i

We denote the complex number $(1, 0)$ by 1. Notice that if (x, y) is any other complex number, then

$$(1,0) \cdot (x,y) = (1 \cdot x - 0 \cdot y, 1 \cdot y + 0 \cdot x) = (x,y).$$

Thus $1 \equiv (1,0)$ is the multiplicative identity. We commonly denote the complex number $(0,1)$ by i. Observe that

$$i \cdot i = (0,1) \cdot (0,1) = (0 \cdot 0 - 1 \cdot 1, 0 \cdot 1 + 1 \cdot 0) = (-1,0) = -1.$$

Thus i is a bona fide square root of -1, but this property is built into the arithmetic of $\mathbb{C}$; it is not achieved by fiat.

It is common to write the complex number (x, y) as $x \cdot 1 + y \cdot i = x + iy$.

7.5.4 Examples of Complex Arithmetic

Some examples of arithmetic operations on $\mathbb{C}$ are:

$$\begin{aligned}
(3+4i)\cdot(5-7i) &= (3,4)\cdot(5,-7) \\
&= (3\cdot 5 - 4\cdot(-7), 3\cdot(-7) + 4\cdot 5) \\
&= (43,-1) = 43 - i. \\
(3-2i) + (9+6i) &= (3,-2) + (9,6) \\
&= (3+9, -2+6) \\
&= (12,4) = 12 + 4i.
\end{aligned}$$

7.5.5 Algebraic Closure of the Complex Numbers

The most important property of the complex numbers is that of *algebraic closure*: any polynomial $p(z) = a_0 + a_1 z + \cdots + a_{n-1} z^{n-1} + a_n z^n$ with complex coefficients has precisely n complex roots (counting multiplicities). This profound fact is due to Gauss, and he produced five distinct proofs. Today there are several dozen proofs of this, the Fundamental Theorem of Algebra.

7.6 The Quaternions

7.6.1 Algebraic Definition of the Quaternions

The quaternions $\mathbb{H}$ consist of 4-tuples (a, b, c, d) of real numbers. To describe their arithmetic properties, it is convenient for us to identify this 4-tuple with an ordered pair of complex numbers $(a+ib, c+id)$ and then in turn to identify that ordered pair with the matrix

$$\begin{pmatrix} a+ib & c+id \\ -c+id & a-ib \end{pmatrix}.$$

The addition law for the quaternions comes from the addition law for these matrices. The multiplication law for the quaternions comes from the multiplication law for these matrices.

7.6.2 A Basis for the Quaternions

It is common to let

$$\mathbf{1} = (1+i0, 0+i0), \ \mathbf{i} = (0+i, 0+i0),$$
$$\mathbf{j} = (0+0i, 1+i0), \ \mathbf{k} = (0+0i, 0+i).$$

Then it is straightforward to verify that

$$\begin{aligned} \mathbf{i}\cdot\mathbf{j} &= \mathbf{k}, \\ \mathbf{j}\cdot\mathbf{k} &= \mathbf{i}, \\ \mathbf{k}\cdot\mathbf{i} &= \mathbf{j}, \\ \mathbf{i}\cdot\mathbf{i} &= -\mathbf{1}, \\ \mathbf{j}\cdot\mathbf{j} &= -\mathbf{1}, \\ \mathbf{k}\cdot\mathbf{k} &= -\mathbf{1}. \end{aligned}$$

Refer to [WEI] for further details about the quaternions.

7.7 The Cayley Numbers

7.7.1 Algebraic Definition of the Cayley Numbers

A typical Cayley number has the form

$$a + b\mathbf{i}_0 + c\mathbf{i}_1 + d\mathbf{i}_2 + e\mathbf{i}_3 + f\mathbf{i}_4 + g\mathbf{i}_5 + h\mathbf{i}_6.$$

The algebraic rule for the Cayley numbers is that each of the 3-tuples $(\mathbf{i}_0, \mathbf{i}_1, \mathbf{i}_3)$, $(\mathbf{i}_1, \mathbf{i}_2, \mathbf{i}_4)$, $(\mathbf{i}_2, \mathbf{i}_3, \mathbf{i}_5)$, $(\mathbf{i}_3, \mathbf{i}_4, \mathbf{i}_6)$, $(\mathbf{i}_4, \mathbf{i}_5, \mathbf{i}_0)$, $(\mathbf{i}_5, \mathbf{i}_6, \mathbf{i}_1)$, $(\mathbf{i}_6, \mathbf{i}_0, \mathbf{i}_2)$ behaves like the triple $(\mathbf{i}, \mathbf{j}, \mathbf{k})$ in the quaternions. We refer the reader to [WEI] for further details on the Cayley numbers.

7.8 Nonstandard Analysis

7.8.1 The Need for Nonstandard Numbers

Isaac Newton's calculus was premised on the existence of certain "infinitesimal numbers"—numbers that are positive, smaller than any standard real number, but not zero. Because limits were not understood in Newton's time, infinitesimals served in their stead. But in fact it was just these infinitesimals that called the theory of calculus into doubt. More than a century was expended developing the theory of limits in order to dispel those doubts.

Nonstandard analysis, due to Abraham Robinson (1918–1974), is a model for the real numbers (i.e., it is a number system that satisfies the axioms for the real numbers that we enunciated in Subsection 7.4.2) that contains infinitesimals. In a sense, then, Robinson's nonstandard reals are a perfectly rigorous theory that vindicates Newton's original ideas about infinitesimally small numbers.

7.8.2 Filters and Ultrafilters

One of the most standard constructions of the nonstandard real numbers involves putting an equivalence relation on the set of all sequences $\{a_j\}$ of real numbers. A natural algebraic construction for doing so is the *ultrafilter*. In fact, ultrafilters are widely used in model theory (see the article by P. C. Eklof in [BAR]). So we shall briefly say now what an ultrafilter is.

Let I be a nonempty set. A *filter* over I is a set $D \subseteq \mathcal{P}(I)$ such that

1. $\emptyset \notin D$, $I \in D$;
2. If $X, Y \in D$, then $X \cap Y \in D$;
3. If $X \in D$ and $X \subseteq Y \subseteq I$, then $Y \in D$.

In particular, a filter D over I has the *finite intersection property*: the intersection of any finite set of elements of D is nonempty.

A filter D over I is called an *ultrafilter* if, for every $X \subseteq I$, either $X \in D$ or $I \backslash X \in D$. It turns out that a filter over I is an ultrafilter if and only if it is a maximal filter over I (i.e., there is no larger filter containing it). One can show, using Zorn's lemma, that if S is a collection of subsets of I that has the finite intersection property, then S is contained in an ultrafilter over I.

7.8.3 A Useful Measure

We will follow the exposition that may be found at

`http://members.tripod.com/PhilipApps/howto.html.`

See also [LIN], [CUT], [LOW], and [DAV1]. At the end, we will point out the ultrafilter that is lurking in the background.

Let m be a finitely additive measure on the set $\mathbb{N}$ of natural numbers such that

1. For any subset $A \subseteq \mathbb{N}$, $m(A)$ is either 0 or 1.

2. It holds that $m(\mathbb{N}) = 1$ and $m(B) = 0$ for any finite set B.

That such a measure m exists is an easy exercise with the Axiom of Choice. We leave the details to the interested reader.

7.8.4 An Equivalence Relation

Let

$$S = \left\{ \{a_n\}_{n=1}^{\infty} : a_n \in \mathbb{R} \text{ for all } n = 1, 2, \ldots \right\}.$$

Define a relation $\sim$ on S by

$$\{a_n\} \sim \{b_n\} \quad \text{if and only if} \quad m\{n : a_n = b_n\} = 1.$$

Then $\sim$ is clearly an equivalence relation. We let $\mathbb{R}^* = S/\sim$ be the nonstandard real number system.[1]

We let $[\{a_n\}]$ denote the equivalence class containing the sequence $\{a_n\}$. Then we define some of the elementary operations on $\mathbb{R}^*$ by

$$[\{a_n\}] + [\{b_n\}] = [\{a_n + b_n\}];$$

$$[\{a_n\}] \cdot [\{b_n\}] = [\{a_n \cdot b_n\}];$$

$$[\{a_n\}] < [\{b_n\}] \quad \text{iff} \quad m(\{n : a_n < b_n\}) = 1.$$

Further, we identify a standard real number b with the equivalence class $[\{b, b, b, \ldots\}]$.

7.8.5 An Extension of the Real Number System

We have seen that $\mathbb{R}^*$ clearly contains $\mathbb{R}$ in a natural way. And it contains other elements too. We call $x \in \mathbb{R}^*$ an *infinitesimal* if and only if $a \neq 0$ and $-a < x < a$ for every positive real number a. For example, $[\{1, 2/3, 1/3, \ldots\}]$ is an infinitesimal. We call $y \in \mathbb{R}^*$ an *infinitary number* if $y > b$ for every real number b or $y < d$ for every real number d. As an instance, $[\{1, 2, 3, \ldots\}]$ is an infinitary number.

It would be inappropriate in a book of this type to delve very far into the theory of the nonstandard reals (see [NEL], [LOW], [LIN], [DAV1], and [CUT] for further details). But at least now the reader has an idea of what the nonstandard real numbers are, and how a number system could contain both the standard reals and also infinitesimals and infinitaries.

[1] In fact, this is the point where we use an ultrafilter. The set $\mathcal{M} = \{A \subseteq \mathbb{N} : m(A) = 1\}$ is an ultrafilter. We are mod-ing out by this ultrafilter.

Chapter 8

Methods of Mathematical Proof

The perfection of mathematical beauty is such ... that whatsoever is most beautiful and regular is also found to be the most useful and excellent.

—Sir D'Arcy Wentworth Thompson

Logical consequences are the scarecrows of fools and the beacons of wise men.

—T.H. Huxley

A system is nothing more than the subordination of all aspects of the universe to any one such aspect.

—Jorge Luis Borges

I do not know what I may appear to the world, but to myself I seem to have been only like a boy playing on the seashore, and diverting myself in now and then finding a smoother pebble or a prettier shell than ordinary, whilst the great ocean of truth lay all undiscovered before me.

—Isaac Newton

The average Ph.D. thesis is nothing but the transference of bones from one graveyard to another.

—J.F. Dobie

Men fear thought as they fear nothing else on earth—more than ruin, more even than death.

—Bertrand Russell

8.1 Axiomatics

8.1.1 Undefinables

The basic elements of mathematics are "undefinables" (Subsection 3.1.1). Since every new piece of terminology is defined in terms of old pieces of terminology, we must begin with certain terms that have no definition.

Most commonly, the terms "set" and "element of" are taken to be undefinables. We simply say that a set S is a collection of objects and x is an element of S if it is one of those objects.

8.1.2 Definitions

From this beginning, we formulate more complex definitions. For example, if A and B are sets, then we can define $A \times B$ to be all ordered pairs (a, b) such that $a \in A$ and $b \in B$. Of course, this presupposes that we have defined $\in$ ("element of") and "ordered pair." Then we can define a function from A to B to be a certain type of subset of $A \times B$. And so forth.

8.1.3 Axioms

Once a collection of definitions is put in place, then we can formulate axioms. An *axiom* is a statement whose truth we take as given. The axiom uses terminology that consists of undefinables plus terms introduced in the definitions. An axiom usually has a subject, a verb, and an object. For example, a famous axiom from Euclidean geometry says

> For each line ℓ and each point $\mathbf{P}$ that does not lie on ℓ there is a unique line m through $\mathbf{P}$ such that m is parallel to ℓ.

Following the spirit of Occam's Razor, we generally endeavor to have as few axioms as possible. Euclidean geometry has just five axioms. Group theory has three axioms, and field theory has eleven axioms. The natural numbers have five axioms. There are eight axioms for the real numbers.

8.1.4 Theorems, *ModusPonendoPonens*, and *ModusTollens*

Next, we begin to formulate theorems. A theorem is a statement that we derive from the axioms using rules of logic. There is really only one fundamental rule of logic, and it is this:

> *modus ponendo ponens*: If $\mathbf{A}$ and $[\mathbf{A} \Rightarrow \mathbf{B}]$, then $\mathbf{B}$.

Although the terminology is less frequently encountered in the literature, some books refer to *modus tollens*. It is the contrapositive form of *modus ponendo ponens*:

> If $\mathbf{B}$ and $[\sim \mathbf{A} \Rightarrow \sim \mathbf{B}]$, then $\mathbf{A}$.

Even though there is only one rule of logic, there are several different proof strategies. The purpose of this chapter is to enunciate, discuss, and illustrate some of the most prominent and useful of these.

8.2 Proof by Induction

The word "induction" is used in ordinary parlance to describe any method of inference. In mathematics it has a very specific meaning, which is summarized as follows.

8.2.1 Mathematical Induction

Mathematical Induction: For each $n \in \mathbb{N}$, let $P(n)$ be a statement. If

(1) $P(1)$ is true;

(2) $P(n) \Rightarrow P(n+1)$ for every natural number n;

then $P(n)$ is true for every n.

The method of induction is best understood by way of several examples. See also Subsections 5.8.11 and 5.8.12.

8.2.2 Examples of Inductive Proof

Example 8.1

Prove that, if n is a positive integer, then

$$1 + 2 + \cdots + n = \frac{n(n+1)}{2}.$$

Proof: Let $P(n)$ be the statement

$$1 + 2 + \cdots + n = \frac{n(n+1)}{2}.$$

Then $P(1)$ is the simple equation

$$1 = \frac{1 \cdot 2}{2}.$$

This is certainly true, so we have established step (1) of the induction process.

The second step is the more subtle. We *assume* $P(n)$, which is

$$1 + 2 + \cdots + n = \frac{n(n+1)}{2}, \qquad (*_n)$$

and we use it to *prove* $P(n+1)$, which is

$$1 + 2 + \cdots + (n+1) = \frac{(n+1)(n+2)}{2}. \qquad (*_{n+1})$$

To accomplish this goal, we add the quantity $(n+1)$ to both sides of $(*_n)$. Thus we have

$$[1 + 2 + \cdots + n] + (n+1) = \frac{n(n+1)}{2} + (n+1). \qquad (\dagger)$$

The right-hand side can be simplified as follows:

$$\begin{aligned}\frac{n(n+1)}{2} + (n+1) &= \frac{n^2 + n + (2n+2)}{2} \\ &= \frac{n^2 + 3n + 2}{2} = \frac{(n+1)(n+2)}{2}.\end{aligned}$$

As a result of this calculation, we may rewrite (†) as

$$1 + 2 + \cdots + (n+1) = \frac{(n+1)(n+2)}{2}.$$

But this is precisely $(*_{n+1})$.

We have assumed $P(n)$ and used it to prove $P(n+1)$. That is step (2) of the induction method. Our proof is complete.

□

Example 8.2

Prove that, for any positive integer n, the quantity $n^2 + 3n + 2$ is even.

Proof: Obviously the statement $P(n)$ must be

The quantity $n^2 + 3n + 2$ is even.

Observe that $P(1)$ is the assertion that $1^2 + 3 \cdot 1 + 2 = 6$ is even. That is obviously true.

Now assume $P(n)$ (i.e., that n^2+3n+2 is even). We must use this hypothesis to prove $P(n+1)$ (i.e., that $(n+1)^2+3(n+1)+2$ is even). Now

$$\begin{aligned}(n+1)^2 + 3(n+1) + 2 &= n^2 + 2n + 1 + 3n + 3 + 2 \\ &= [n^2 + 3n + 2] + [2n + 4] \\ &= [n^2 + 3n + 2] + 2[n + 2].\end{aligned}$$

The number $n^2 + 3n + 2$ is even by the inductive hypothesis $P(n)$. And $2[n + 2]$ is even since it is a multiple of 2. The sum of two even numbers is even, because each will be a multiple of 2. So we see that $(n+1)^2+3(n+1)+2$ is even. That establishes $P(n+1)$, assuming $P(n)$. The inductive proof is complete.

□

The next example illustrates a mathematical device, due to Peter Gustav Lejeune Dirichlet (1805–1859), known as the the pigeonhole principle. In early days it was known as the *Dirichletscher Schubfachschluss.*

Example 8.3

Prove that if $n+1$ letters are placed in n mailboxes, then some mailbox will contain (at least) two letters.

Proof: Let $P(n)$ be the statement

If $n+1$ letters are put in n mailboxes, then some mailbox will contain (at least) two letters.

Then $P(1)$ is the simple assertion that if two letters are placed in one mailbox, then some mailbox contains at least two letters. This is trivial: there is one mailbox and it indeed contains two letters.

Now we suppose that $P(n)$ is true and we use that statement to prove $P(n+1)$. Now suppose that $n+2$ letters are placed into $n+1$ mailboxes. There are three possibilities:

- If the last mailbox contains no letter, then all of the letters actually go into the first n mailboxes. So the inductive hypothesis $P(n)$ applies. Therefore, some mailbox contains at least two letters.
- If the last mailbox contains only one letter, then $n+1$ letters have gone into the first n mailboxes, and the inductive hypothesis $P(n)$ applies. So some mailbox contains at least two letters.
- If the last mailbox contains (at least) two letters, then we have identified a box with two letters.

Thus, by breaking the proof into three cases, we have established $P(n+1)$ (assuming $P(n)$). The proof is complete.

□

Example 8.4

Let us prove that a set with n elements has 2^n subsets.

Proof: Our inductive statement is

$\boldsymbol{P(n)}$: A set with n elements has 2^n subsets.

Now $P(1)$ is clearly true: A set with 1 element has $2 = 2^1$ subsets, namely the empty set and the set itself.

Suppose inductively that $P(n)$ has been established. Consider now a set $A = \{a_1, \ldots, a_{n+1}\}$ with $n + 1$ elements. Write $A = \{a_1, \ldots, a_n\} \cup \{a_{n+1}\} \equiv A' \cup \{a_{n+1}\}$. Now A' is a set with n elements, so the inductive hypothesis applies to it. The set A' therefore has 2^n subsets. These are also, of course, subsets of A. The additional subsets of A are obtained by adjoining the element a_{n+1} to each of the subsets of A'. That gives 2^n more subsets of A, for a total of $2^n + 2^n$ subsets.

We conclude that A has $2^{n+1} = 2^n + 2^n$ subsets. That completes the inductive step, and the proof.

□

8.2.3 Complete or Strong Mathematical Induction

We conclude this section by mentioning an alternative form of the induction paradigm, which is sometimes called *complete mathematical induction* or *strong mathematical induction.*

> **Complete Mathematical Induction**: Let **P** be a function on the natural numbers. If
>
> 1. $P(1)$;
> 2. $[P(j) \text{ for all } j \leq n] \Rightarrow P(n + 1)$ for every natural number n;
>
> then $P(n)$ is true for every n.

It turns out that the complete induction principle is logically equivalent to the ordinary induction principle enunciated at the outset of this section. But in some instances strong induction is the more useful tool. Alternative terminologies for complete induction are "the set formulation of induction," "total induction," and "course-of-values induction."

Complete induction is sometimes more convenient, or more natural, to use than ordinary induction; it finds particular use in abstract algebra. Complete induction also is a simple instance of transfinite induction (see Section 5.8.12).

Example 8.5

Theorem: Every integer greater than 1 is either prime or the product of primes. (Here, a prime number is an integer whose only factors are 1 and itself.)

Proof: We will use strong induction, just to illustrate the idea. For convenience, we begin the induction process at the index 2 rather than at 1.

Let $P(n)$ be the assertion "Either n is prime or n is the product of primes." Then $P(2)$ is plainly true because 2 is the first prime. Now assume that $P(k)$ is true for $2 \leq k \leq n$ and consider $P(n+1)$. If $n+1$ is prime, then we are done. If $n+1$ is not prime, then $n+1$ factors as $n+1 = k \cdot \ell$, where k, ℓ are integers less than $n+1$, but at least 2. By the strong inductive hypothesis, each of k and ℓ factors as a product of primes (or is itself a prime). Thus $n+1$ factors as a product of primes.

The complete induction is done, and the proof is complete.

□

8.3 Proof by Contradiction

Proof by contradiction is predicated on the classical "law of the excluded middle"—an idea that goes back to Aristotle.[1] The substance of the proof strategy is that an idea is either true or false (see Subsections 1.2.1 and 1.2.2 on truth values). There is no "middle" status. With this premise in mind, we can prove that something is true by excluding the possibility that it is false. The way that we exclude the possibility that the assertion is false is to assume it is false and show that such an assumption leads to an untenable position (i.e., a contradiction). The only possible conclusion therefore is that the assertion is true. We now illustrate with some examples.

8.3.1 Examples of Proof by Contradiction

Example 8.6

Prove that if $n+1$ letters are placed in n mailboxes, then one mailbox must contain (at least) two letters.

Proof: Seeking a contradiction, we suppose the contrary. Thus we have a way to put $n+1$ letters into n mailboxes so that each mailbox contains only 0 or 1 letter. Let m_j be the number of letters in the jth mailbox. Then

$$n+1 = \sum_{j=1}^{n} m_j \leq \sum_{j=1}^{n} 1 = n.$$

[1]The law of the excluded middle is sometimes referred to as *tertium non datur*.

Under our hypothesis, we have derived the absurd statement that $n + 1 \leq n$. That is a contradiction. As a result, our hypothesis must be false, and some mailbox must contain (at least) two letters.

□

Example 8.7

Prove that if n is a positive integer, then $n^2 + 3n + 2$ is even.

Proof: If not, then $n^2 + 3n + 2$ is odd for some n. Any odd number has the form $2m + 1$ for some integer m. Hence

$$n^2 + 3n + 2 = 2m + 1.$$

But then

$$n^2 + 3n - 2m = -1,$$

or

$$n(n + 3) - 2m = -1.$$

Now, if n is even, then $n+3$ is odd and if n is odd, then $n+3$ is even. In either case, $n(n + 3)$ will be the product of an even and an odd number and will thus be even. So $n(n+3) = 2k$ for some integer k. As a result we have

$$2k - 2m = -1$$

or

$$2(k - m) = -1.$$

But this shows that the number -1 is even, and that is impossible. We conclude that our initial hypothesis is false: n^2+2n+3 cannot be odd; it must be even.

□

Example 8.8

Theorem (Pythagoras): There is no rational number whose square is 2.

Proof: To the contrary, suppose that there is such a rational number α. We write $\alpha = p/q$, expressing the fraction in lowest terms (i.e., p and q have no common prime factors).

Then

$$2 = \alpha^2 = p^2/q^2.$$

Clearing denominators, we see that

$$2q^2 = p^2.$$

But 2 divides the left-hand side and therefore 2 divides the right-hand side. It follows that 2 divides p. Write $p = 2r$. Then

$$2q^2 = [2r]^2$$

or

$$q^2 = 2r^2.$$

Now 2 divides the right-hand side; therefore 2 divides the left-hand side. Thus 2 divides q. But now we have established both that 2 divides p and that 2 divides q. This assertion contradicts the hypothesis that p and q have no common factors.

We conclude that α cannot exist.

□

8.4 Direct Proof

A direct proof is one in which a sequence of logical steps leading ever closer to the desired conclusion is produced. There are no additional logical tricks, such as induction or proof by contradiction. The concept is best illustrated through examples.

8.4.1 Examples of Direct Proof

Example 8.9

Prove that, if n is a positive integer, then the quantity n^2+3n+2 is even.

Proof: Denote the quantity $n^2 + 3n + 2$ by K. Observe that

$$K = n^2 + 3n + 2 = (n + 1)(n + 2).$$

Thus K is the product of two successive integers: $n+1$ and $n+2$. One of those two integers must be even. So it is a multiple of 2. Therefore K itself is a multiple of 2. Hence, K must be even.

□

Example 8.10

Prove that the sum of an even integer and an odd integer is odd.

Proof: An even integer is divisible by 2, so it may be written in the form $e = 2m$, where m is an integer. An odd integer has remainder 1 when divided by 2, so it may be written in the form $o = 2k + 1$, where k is an integer. The sum of these is

$$e + o = 2m + (2k + 1) = 2(m + k) + 1.$$

Thus we see that the sum of an even and an odd integer will have remainder 1 when it is divided by 2. As a result, the sum is odd.

□

Example 8.11

Prove that every even integer may be written as the sum of two odd integers.

Proof: Let the even integer be $K = 2m$, for m an integer. If m is odd, then we write

$$K = 2m = m + m$$

and we have written K as the sum of two odd integers. If, instead, m is even, then we write

$$K = 2m = (m - 1) + (m + 1).$$

Since m is even, then both $m - 1$ and $m + 1$ are odd. So again we have written K as the sum of two odd integers.

□

Example 8.12

Prove the Pythagorean theorem.

Proof: The Pythagorean theorem states that $c^2 = a^2 + b^2$, where a and b are the legs of a right triangle and c is its hypotenuse (see Figure 8.1). Consider now the arrangement of four triangles and a square shown in Figure 8.2. We assume that $b > a$. Each of the four triangles is a copy of the original triangle in Figure 8.1. We see that each side of the all-encompassing square is equal to c, so the area of that square is c^2. Now each of the component triangles has base a and height b. Hence each such triangle has area $ab/2$. And the little square in the middle has side $b - a$, so it has area $(b - a)^2 = b^2 - 2ab + a^2$. We write the total area as the sum of its component areas:

$$c^2 = 4 \cdot \left[\frac{ab}{2}\right] + [b^2 - 2ab + a^2] = a^2 + b^2.$$

That is the desired equality.

□

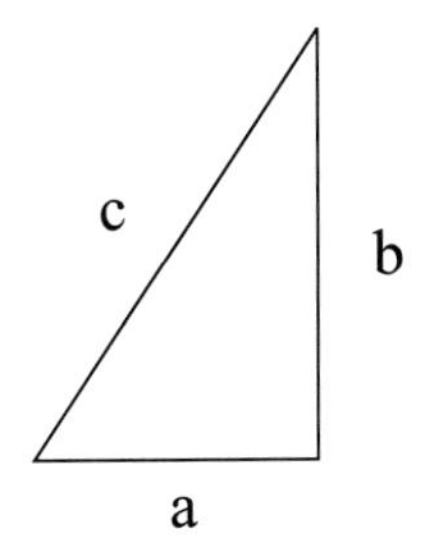

Figure 8.1

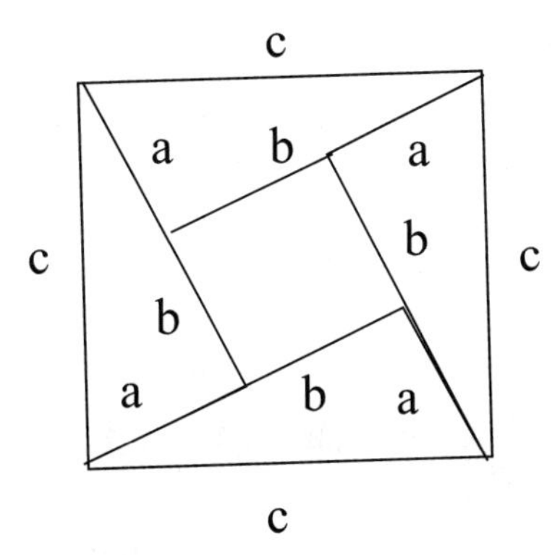

Figure 8.2

8.5 Other Methods of Proof

Induction, contradiction, and direct proof are the three most common proof techniques. Almost any proof can be shoehorned into one of these three paradigms. But there are other techniques that should be mentioned. One of these is enumeration, or counting. We illustrate this method with some examples.

8.5.1 Examples of Counting Arguments

Example 8.13

Show that if there are 23 people in a room, then the odds are better than even that two of them have the same birthday.

Proof: The best strategy is to calculate the odds that *no two* of the people have the same birthday, and then to take complements.

Let us label the people $p_1, p_2, \ldots, p_{23}$. Then, assuming that none of the p_j have the same birthday, we see that p_1 can have a birthday on any of the 365 days in the year, p_2 can then have a birthday on any of the remaining 364 days, p_3 can have a birthday on any of the remaining 363 days, and so forth. Thus the number of different ways that 23 people can all have different birthdays is

$$365 \cdot 364 \cdot 363 \cdots 345 \cdot 344 \cdot 343.$$

On the other hand, the number of ways that birthdays could be distributed (with no restrictions) among 23 people is

$$\underbrace{365 \cdot 365 \cdot 365 \cdots 365}_{23 \text{ times}} = 365^{23}.$$

Thus, the probability that 23 people all have different birthdays is

$$p = \frac{365 \cdot 364 \cdot 363 \cdots 343}{365^{23}}.$$

A quick calculation with a pocket calculator shows that $p \sim 0.4927 < .5$. That is the desired result.

□

Example 8.14

Show that if there are six people in a room, then either three of them know each other or three of them do not know each other.

(Here three people know each other if each of the three pairs has met. Three people do not know each other if each of the three pairs has *not* met.)

Proof: The tedious way to do this problem is to write out all possible "acquaintance assignments" for 15 people.

We now describe a more efficient, and more satisfying, strategy. Call one of the people Bob. There are five others. Either Bob knows three of them, or he does not know three of them.

Say that Bob knows three of the others. If any two of those three are acquainted, then those two and Bob form a mutually acquainted threesome. If no two of those three know each other, then those three are a mutually unacquainted threesome.

Now suppose that Bob does not know three of the others. If any two of those three are unacquainted, then those two and Bob form an unacquainted threesome. If all pairs among the three are instead acquainted, then those three form a mutually acquainted threesome.

We have covered all possibilities, and in every instance come up either with a mutually acquainted threesome or a mutually unacquainted threesome. That ends the proof.

□

It may be worth noting that five people is insufficient to guarantee either a mutually acquainted threesome or a mutually unacquainted threesome. We leave it to the reader to provide a suitable counterexample. It is quite difficult to determine the minimal number of people to solve the problem when "threesome" is replaced by "foursome." When "foursome" is replaced by five people, the problem is considered to be grossly intractable. This problem is a simple example from the mathematical subject known as Ramsey theory.

Example 8.15

Jill is dealt a poker hand of five cards from a standard deck of 52. What is the probability that she holds four of a kind?

Proof: If the hand holds four aces, then the fifth card is any one of the other 48 cards. If the hand holds four kings, then the fifth card is any one of the other 48 cards. And so forth. So there are a total of

$$13 \times 48 = 624$$

possible hands with four of a kind. The total number of possible five-card hands is

$$\binom{52}{5} = 2598960.$$

Therefore, the probability of holding four of a kind is

$$p = \frac{624}{2598960} = 0.00024.$$

□

The last example is not quite a proof by contradiction and not quite a proof by exhaustion.

Example 8.16

Let us show that there exist irrational numbers a and b such that a^b is rational.

Let $\alpha = \sqrt{2}$ and $\beta = \sqrt{2}$. If α^β is rational, then we are done, using $a = \alpha$ and $b = \beta$. If α^β is irrational, then observe that

$$\left(\alpha^\beta\right)^{\sqrt{2}} = \alpha^{[\beta\cdot\sqrt{2}]} = \alpha^2 = [\sqrt{2}]^2 = 2.$$

Thus, with $a = \alpha^\beta$ and $b = \sqrt{2}$ we have found two irrational numbers a, b such that $a^b = 2$ is rational.

□

Curiously, in this last example, we are unable to say which two irrational numbers do the job. But we have proved that two such numbers exist.

Chapter 9

The Axiom of Choice

To choose one sock from each of infinitely many pairs of socks requires the Axiom of Choice, but for shoes the Axiom is not needed.

—Bertrand Russell

The Axiom of Choice is obviously true; the well-ordering principle is obviously false; and who can tell about Zorn's lemma?

—Jerry Bona

The test of every religious, political, or educational system, is the man which it forms. If a system injures the intelligence it is bad. If it injures the character it is vicious, if it injures the conscience it is criminal.

—Henri Frédéric Amiel

Every person thinks his own intellect perfect, and his own child handsome.

—Sa'di

The Axiom of Choice allows you to take things out of sets that you never should have been allowed to put in there in the first place.

—anon.

A choice function exists in constructive mathematics, because a choice is implied by the very meaning of existence.

—Errett Bishop

Thought is subversive and revolutionary, destructive and terrible; thought is merciless to privilege, established institutions, and comfortable habit. Thought looks into the pit of hell and is not afraid. Thought is great and swift and free, the light of the world, and the chief glory of man.

—Bertrand Russell

It is no exaggeration to say that a straightforward realistic approach to mathematics has yet to be tried. It is time to make the attempt.

—Errett Bishop

9.1 Enunciation of the Axiom

The Axiom of Choice was first enunciated by Zermelo. The standard formulation is as follows.

Let S be a set. Then there is a function

$$f : \mathcal{P}(S) \to S$$

such that $f(A) \in A$ for each nonempty $A \in \mathcal{P}(S)$.

In English, the Axiom of Choice states that there is a way to select an element from each subset of S. In logic, the name of this axiom is frequently abbreviated AC.

If $S = \mathbb{N}$ (the natural numbers), then the validity of the Axiom of Choice is obvious: for any $A \subset S$, let $f(A)$ be the least element of A. A similar observation applies to any well-ordered set S. By contrast, it is known to be impossible to write the choice function explicitly when $S = \mathbb{R}$.

9.2 Examples of the Use of the Axiom of Choice

The book [JEC] is an authoritative source for information about the Axiom of Choice. See also [HRJ] for context. In this section, we illustrate the importance of this axiom in modern mathematics by describing some of its most common occurrences. A more exhaustive treatment of equivalents of the Axiom of Choice appears in [RR1] and [RR2].

9.2.1 Zorn's Lemma

In modern mathematics, especially in algebra, Zorn's lemma plays a central role. It is used to prove the existence of maximal ideals, of bases for vector spaces, and of other "maximal sets."

We need two pieces of terminology to formulate Zorn's lemma. First, if $(S, \leq)$ is a partially ordered set, then a *chain* in S is a subset $C \subseteq S$ that is linearly ordered (i.e., any two elements are comparable). An element u is an upper bound of the chain C if $c \leq u$ for every $c \in C$.

A typical enunciation of Zorn's lemma is this:

Let $(S, \leq)$ be a nonempty, partially ordered set with the property that every chain in S has an upper bound. Then S has a maximal element (i.e., an element x such that $s \leq x$ for every $s \in S$).

Zorn's lemma is equivalent to the Axiom of Choice.

9.2.2 The Hausdorff Maximality Principle

The Hausdorff maximality principle is a variant of Zorn's lemma, also commonly used in algebraic applications.

> If $\mathcal{R}$ is a transitive relation on a set S, then there exists a maximal subset of S that is linearly ordered by $\mathcal{R}$.

Hausdorff's principle is equivalent to the Axiom of Choice.

9.2.3 The Tukey–Tychanoff Lemma

Yet another variant of the Zorn's lemma/Hausdorff maximality principle is the so-called Tukey–Tychanoff lemma. We say in what follows that a family of sets $\mathcal{F}$ has *finite character* if, for each set X, X belongs to $\mathcal{F}$ if and only if every finite subset of $\mathcal{F}$ belongs to $\mathcal{F}$. Now the Tukey–Tychanoff lemma is formulated in this way:

> Let $\mathcal{F}$ be a nonempty family of sets. If $\mathcal{F}$ has finite character, then $\mathcal{F}$ has a maximal element (with respect to the partial ordering $\subseteq$).

The Tukey–Tychanoff lemma is equivalent to the Axiom of Choice.

9.2.4 A Maximum Principle for Classes

Recall that classes (Subsection 5.9.2) are collections of objects that are (in principle) larger than sets, but in a very specifically allowable manner. A *nest* is a class that is linearly ordered by inclusion. Now we have

> If every nonempty nest that is a subset of a nonempty class X has the property that its union is an element of X, then either X has a maximal element or there is a subclass of X that is a nest and a proper class.

This result, too, is equivalent to the Axiom of Choice.

The books [RR1] and [RR2] contain several dozen additional maximum principles that are either equivalent to, or closely related to, various versions of the Axiom of Choice.

9.3 Consequences of the Axiom of Choice

Although the brevity of this volume precludes an exploration of the proofs, we can still provide a considerable variety of results here—from various branches of mathematics—to illustrate the broad utility of the Axiom of Choice. Many of these results are equivalent to some version of the Axiom of Choice, others are implied by one version and imply another, and yet others are only consequences of the Axiom of Choice. We refer the reader to [JEC], [RR1], and [RR2] for a complete discussion of this taxonomy.

Theorem 9.1 (a nonmeasurable set)
There is a bounded set $E \subseteq [0,1] \subseteq \mathbb{R}$ that is not Lebesgue measurable in the sense that it is impossible to assign a finite value as the Lebesgue measure of E.

Theorem 9.2 (Tychanoff's product theorem)
The product of (any number of) compact sets is compact in the product topology.

Theorem 9.3 (product of nonempty sets)
The product of (any number of) nonempty sets is nonempty.

Theorem 9.4 (existence of maximal ideals)
Let R be a ring and J an ideal in R. Then there is a maximal ideal M that contains J.

Theorem 9.5 (existence of a basis for a vector space)
Let V be any vector space over a field k. Then there exists a basis for V.

Theorem 9.6 (Nielsen–Schreier)
Every subgroup of a free group is a free group.

Let V be a real vector space. A functional p on E is said to be *sublinear* if $p(x+y) \leq p(x)+p(y)$ for all $x, y \in V$ and also $p(rx) = rp(x)$ for all nonnegative real numbers r and all $x \in V$.

Theorem 9.7 (Hahn–Banach)
Let p be a sublinear functional on V, and let ϕ be a linear functional defined on a subspace $W \subseteq V$ such that $\phi(x) \leq p(x)$ for all $x \in W$. Then there is a linear functional ψ on V that extends ϕ and so that $\psi(x) \leq p(x)$ for all $x \in V$.

Theorem 9.8 (existence of the algebraic closure)
Let F be a field. Then the algebraic closure of F exists and is unique up to isomorphism.

Theorem 9.9 (prime ideal theorem)
Every Boolean algebra has a prime ideal.

Theorem 9.10 (compactness theorem of logic)
In first-order logic, if every finite subset of Σ has a model, then Σ has a model.

Theorem 9.11 (trichotomy)
For any two sets A and B, either there is an injection of A into B or there is an injection of B into A.

Theorem 9.12 (well-ordering principle)
Any set S can be well-ordered.

Theorem 9.13 (Artin–Schreier)
Any field in which -1 is not a sum of squares can be ordered.

Notice that this last theorem certainly excludes the complex numbers—a field that cannot be ordered.

9.4 Paradoxes

There are some startling paradoxes connected with either the Axiom of Choice or its denial. It is known [SOL] that if the Axiom of Choice is denied, then there is a model for the real numbers in which all sets are measurable. It is also known [WRI] that if the Axiom of Choice is denied, then there is a model for Hilbert space in which all operators are bounded.

Perhaps even more surprising is the following Banach–Tarski paradox. We will formulate just one version of the phenomenon. An authoritative treatment appears in [JEC].

> Let B be the closed unit ball in $\mathbb{R}^3$. There is a decomposition
>
> $$B = S \cup T,$$
>
> with $S \cap T = \emptyset$, such that $S = S_1 \cup S_2 \cup \cdots \cup S_k$, $T = T_1 \cup T_2 \cup \ldots \cup T_k$, $B = B_1 \cup B_2 \cup \cdots \cup B_k$ and, for each j, S_j, T_j, and B_j are geometrically congruent.

In simple language, the Banach–Tarski paradox says that the closed unit ball may be broken up into finitely many pieces that can be reassembled into two disjoint closed unit balls.

9.5 The Countable Axiom of Choice

The Countable Axiom of Choice asserts that every countable set has a choice function. This statement is strictly weaker than the full Axiom of Choice.

The Countable Axiom of Choice can be used to prove a number of useful results in basic analysis. For example, it can be used to show that the countable union of countable sets is countable. Thus the real number system is not a countable union of countable sets. The Countable Axiom of Choice shows that any subspace of a separable metric space is separable. The Countable Axiom of Choice shows that Lebesgue measure is countably additive.

An interesting consequence of the Countable Axiom of Choice (by way of the Principle of Dependent Choices—see [JEC, pp. 20 ff, 119]) is the following characterization of well-orderings:

> A linear ordering $>$ of a set S is a well-ordering if and only if S has no infinite descending sequence:
>
> $$x_0 > x_1 > x_2 \cdots > x_n > \cdots .$$

9.6 Consistency of the Axiom of Choice

Gödel's method of establishing the consistency of the Axiom of Choice with the other axioms of set theory is by way of constructing a model. In particular, he created a model of set theory called "the constructible sets," which in effect is all the sets that can be described with first-order language using transfinite induction. An alternative construction is that of the "definable sets," namely all the sets that are hereditarily ordinal-definable. In Gödel's model of set theory, all of the usual axioms of set theory hold and AC holds as well.

9.7 Independence of the Axiom of Choice

In 1963, Paul Cohen used the method of forcing to construct a model of set theory in which the Axiom of Choice is false. This result, together with Gödel's consistency theorem, establishes that the Axiom of Choice is independent of the other axioms of set theory.

Chapter 10

Proof Theory

Proofs are the last thing looked for by a truly religious mind which feels the imaginative fitness of its faith.

—George Santayana

Faith embraces many truths which seem to contradict each other.

—Blaise Pascal

Imagination is more robust in proportion as reasoning power is weak.

—Giambattista Vico

But if thought is to become the possession of the many, not the privilege of the few, we must have done with fear. It is fear that holds men back—fear lest they should prove less worthy of respect than they have supposed themselves to be.

—Bertrand Russell

Where all is but dream, reasoning and arguments are of no use, truth and knowledge nothing.

—John Locke

All beliefs are demonstrably true. All men are demonstrably in the right. Anything can be demonstrated by logic.

—Antoine de Saint-Exupéry

Logic: The art of thinking and reasoning in strict accordance with the limitations and incapacities of the human understanding.

—Ambrose Bierce

The want of logic annoys. Too much logic bores. Life eludes logic, and everything that logic alone constructs remains artificial and forced.

—André Gide

The seed haunted by the sun never fails to find its way between the stones in the ground. And the pure logician, if no sun draws him forth, remains entangled in his logic.

—Antoine de Saint-Exupéry

10.1 General Remarks

Proof theory is that part of mathematics that studies the concept of proof, the structure of proof, and mathematical provability. A proof theorist is interested in the process of finding a proof, in machine-generated proof, and in finding the shortest proof of a given statement. The proof theorist is also interested in which axiom schemes might generate proofs most efficiently. Following [BUS], we note that for many purposes a very long proof that is easy to find is much more useful than a short proof that is hard to find.

There are two types of proofs: social proofs and formal proofs. Roughly speaking, a social proof is an argument made by one mathematician to convince another mathematician that a certain statement (a "theorem") is true. For a social proof, the "proof is in the pudding": if the target mathematician is convinced by the argument, then the social proof is acceptable. A formal proof is a much more structured and rigorous entity. In formal proof theory, the statement to be proved must be rendered in a carefully structured sequence of symbols, and the proof itself is given by a sequence of symbols. The rules for constructing the formal proof are explicit and rigorous and can be implemented by a machine.

Proof theory as a discipline concerns itself with formal proofs. This is so because a formal proof has a well-defined structure that is amenable to mathematical analysis. The tasks addressed in proof theory are these:

- To formulate systems of logic and sets of axioms that are appropriate for formalizing mathematical proofs and to characterize which mathematical results follow from which collections of axioms. Phrased differently, the proof theorist examines the proof-theoretic strength of certain formal systems.

- To study the structure of formal proofs. One interesting task is to find normal forms for proofs and to establish syntactic paradigms for proofs.

- To determine what sorts of additional information (besides the truth of the theorem under study) can be extracted from a proof. For example, a proof could contain constructive or computational information.

- To find how to construct formal proofs, such as with a computer.

10.2 Cut Elimination

One of the big ideas of proof theory is cut elimination. The motivation is as follows. Fix attention on the propositional calculus. Suppose that we have a statement A and we wish to find a proof for it using a computer.

A natural way to proceed is to reason backward. But suppose that there is only one proof of A (we could not know this in advance, but it is certainly possible), and that proof contains a line of the form

$$[C \Rightarrow (A \vee B)] \wedge [(B \wedge C) \Rightarrow A] \Rightarrow [C \Rightarrow A]. \qquad (*)$$

It is easy to check with a truth table that this line is true. Such a syllogism is called a *cut.*

The trouble is that the line contains the spurious sentence B. The choice of B is arbitrary, and there is no obvious way to reason backward from A and to determine that B is part of the reasoning. The computer would literally have to search through every possible B.

Thus it would be advantageous, for the sake of computer-generated proofs, to know that any statement that can be proved can in fact be proved without using a cut. That is the content of the basic cut elimination theorem. One can even then estimate the number of steps in the cut-free proof.

A delightful and thoroughgoing discussion of cut elimination, including cut elimination for first-order logic, appears in [BUS, pp. 11 ff.].

10.3 Propositional Resolution

We note in passing that a proof system that addresses the desideratum of being amenable to proof search is one based on the idea of "resolution." If C and D are clauses (i.e., disjunctions of propositional variables or their negations) and if $x \in C$ and $\sim x \in D$, then the *resolution rule* applied to C and D is the inference $C \wedge D$ implies $(C \setminus \{x\}) \cup (D \setminus \{\sim x\})$. If Γ is a set of clauses, then a *resolution refutation* of Γ is a sequence $C_1, C_2, \ldots, C_k$ of clauses such that each C_j is either in Γ or is deduced from earlier members of the sequence by the resolution rule. We also require that the last clause C_k be the empty clause.

The basic theorem now follows.

Theorem 10.1 (completeness for propositional resolution)
If Γ is an unsatisfiable set of clauses, then there is a resolution refutation of Γ.

A significant part of the importance of propositional resolution is connected to the fact that it leads to efficient proof methods in first-order logic.

10.4 Interpolation

Let A and B be formulas such that $A \Rightarrow B$ is valid. An *interpolant* for A and B is a third formula C such that $A \Rightarrow C$ and $C \Rightarrow B$ are both valid. Clearly, the idea of interpolation is closely related to the idea of a cut (Section 10.2). It is a fundamental result that it is always possible to find an interpolant C, containing only nonlogical symbols, which symbols are taken from A and B.

To formulate Craig's interpolation theorem, we need some preliminaries. We let $\perp$ denote the false or empty clause, and we let $\top$ denote the true clause (see [BUS, p. 607] for details on these operators). We write $L(A)$ to denote the set of nonlogical symbols occurring in A plus all free variables occurring in A.

Theorem 10.2 (Craig)
Let A and B be first-order formulas such that $\models A \Rightarrow B$. Then there is a formula C such that $L(C) \Rightarrow [L(A) \cap L(B)]$ and such that $\models A \Rightarrow C$ and $\models C \Rightarrow B$.

Craig's theorem is proved using cut elimination. We can say no more about it here.

Lyndon's theorem extends Craig's in this way: we may require that every predicate symbol that occurs positively (resp., negatively) in C also occurs positively (resp., negatively) in both A and B.

10.5 Finite Type

10.5.1 Universes

A *universe* is a pair $\mathcal{U} = (\text{Set}_{\mathcal{U}}, \text{Fn}_{\mathcal{U}})$. Here $\text{Set}_{\mathcal{U}}$ is the sets of $\mathcal{U}$ and $\text{Fn}_{\mathcal{U}}$ is the functions of $\mathcal{U}$. We let $(A \to B)_{\mathcal{U}} \equiv \{f \in \text{Fn}_{\mathcal{U}} | f : A \to B\}$. The *finite types* of $\mathcal{U}$ over sets $A_0, \ldots, A_n$ are the sets generated from $A_0, \ldots A_n$ by closing under the operations

$$A, B \mapsto A \times B,$$

$$(A \to B)_{\mathcal{U}},$$

$$\mathcal{P}(A).$$

Type symbols σ are used to denote members A_σ of $\mathcal{U}$ that are generated in this way. In detail, given symbols $\gamma_0, \gamma_1, \ldots, \gamma_n$ for the *ground types*, new types are built up by formal operations

$$\sigma, \tau \mapsto \sigma \times \tau,$$

$$(\sigma \to \tau),$$
$$[\sigma].$$

Then

$$\begin{aligned} A_{\gamma_i} &\equiv A_i, \\ A_{\sigma \times \tau} &\equiv A_\sigma \times A_\tau, \\ A_{(\sigma \to \tau)} &\equiv (A_\sigma \to A_\tau), \\ A_{[\sigma]} &\equiv \mathcal{P}(A_\sigma). \end{aligned}$$

The universe $\mathcal{U}$ is said to be of *finite type* over $A_0, \ldots, A_n$ if each subset of X belongs to some $A_{[\sigma]}$ (i.e., to a subset of some A_σ).

10.5.2 Conservative Systems

Type theory is useful in proving, for example, that certain systems are conservative. A system S is said to be *conservative* over a subsystem T if any result that can be proved in S can also be proved in T. For instance, the Axiom of Extensionality can be shown to be unnecessary in certain models of arithmetic using type theory. See [BAR, p. 935] for details.

10.6 Beth's Definability Theorem

10.6.1 Introductory Remarks

Let $\mathbf{P}$ and P' be predicate symbols with the same k-arity. Let $\Gamma(P)$ be any set of first-order sentences that do not involve P', and let $\Gamma(P')$ be the very same set of sentences with each occurrence of $\mathbf{P}$ replaced by P'.

We say that the set $\Gamma(P)$ *explicitly defines* the predicate $\mathbf{P}$ if there is a formula $A(\mathbf{c})$ such that

$$\Gamma(P) \vdash \forall \mathbf{x}, (A(\mathbf{x}) \leftrightarrow P(\mathbf{x})).$$

We say that $\Gamma(P)$ *implicitly defines* the predicate $\mathbf{P}$ if

$$\Gamma(P) \cup \Gamma(P') \models \forall \mathbf{x}, (P(\mathbf{x}) \leftrightarrow P'(\mathbf{x})).$$

10.6.2 The Theorem of Beth

Beth's result states that implicit and explicit definability are the same. In effect, implicit definability of a predicate $\mathbf{P}$ is equivalent to the possibility of uniquely characterizing $\mathbf{P}$. So Beth's theorem (Theorem 10.3) states that a statement that can be uniquely characterized can be explicitly defined by a formula that does not involve $\mathbf{P}$.

Theorem 10.3
The set of sentences $\Gamma(P)$ implicitly defines $\mathbf{P}$ *if and only if it explicitly defines* $\mathbf{P}$.

Chapter 11

Category Theory

In the abstracting of an idea one may lose the very intimate humanity of it.

—Ben Shahn

Abstract qualities begin / With capitals always: / The True, the Good, the Beautiful— / Those are the things that pay!

—Lewis Carroll

Insanity is often the logic of an accurate mind overtasked.

—O.W. Holmes

The utmost abstractions are the true weapons with which to control our thought of concrete fact.

—Alfred North Whitehead

Women have no sense of the abstract—a woman admiring the sky is a woman caressing the sky. In a woman's mind beauty is something she needs to touch.

—Jean Giraudoux

The book-worm wraps himself up in his web of verbal generalities, and sees only the glimmering shadows of things reflected from the minds of others.

—William Hazlitt

Particulars, as everyone knows, make for virtue and happiness; generalities are intellectually necessary evils. Not philosophers but fret-sawyers and stamp collectors compose the backbone of society.

—Aldous Huxley

Intellectual generalities are always interesting, but generalities in morals mean absolutely nothing.

—Oscar Wilde

The cause of all human evils is the not being able to apply general principles to special cases.

—Epictetus

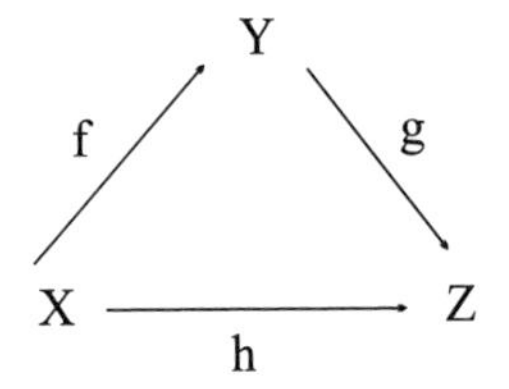

Figure 11.1

11.1 Introductory Remarks

To quote Saunders Mac Lane (inventor, along with S. Eilenberg, of category theory), "Many properties of mathematical constructions may be represented by universal properties of diagrams" (see [MAC]). Category theory begins with the insight that many properties of mathematical systems can be unified and simplified by an illustration with diagrams and arrows. An arrow $f : X \to Y$ represents a function. In particular, a diagram like Figure 11.1 could illustrate a relationship among groups, or topological spaces, or modules, or any number of other mathematical constructs. Category theory elicits certain universal relations, encapsulated in diagrams such as this.

The *lingua franca* of category theory contains the terms *category*, *functor*, *monoid*, *natural transformation*, and *metacategory*. These words describe epistemological abstractions that generalize common mathematical constructs such as ring, homomorphism, group, isomorphism, and mapping. In this brief chapter we describe some of the most basic language of category theory and illustrate it with some simple examples.

11.2 Metacategories and Categories

11.2.1 Metacategories

There are many different ways to develop category theory. Here we follow the exposition in [MAC]. A subject such as category theory begins at the very foundations of mathematics, so a number of definitions are necessary. We take "object" and "arrow" to be undefinables. An object is a mathematical entity on which we operate; an arrow is a function of some kind.

A *metagraph* is a collection of objects $a, b, c, \ldots$, arrows $f, g, h, \ldots$, and two operations:

- The operation of *Domain*, which assigns to each arrow f an object $a = \operatorname{dom} f$;

- The operation of *Codomain*, which assigns to each arrow f an

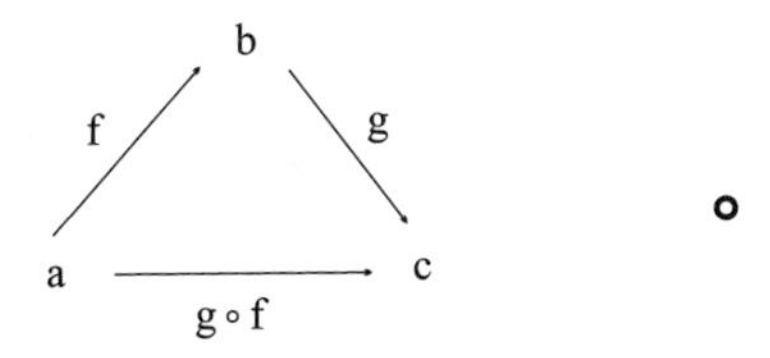

Figure 11.2

object $b = \operatorname{cod} f$.

We typically write

$$f : a \to b \qquad \text{or} \qquad a \xrightarrow{f} b.$$

A *metacategory* is a metagraph with two additional operations:

- The operation of *identity*, which assigns to each object a an arrow $\mathrm{id}_a = 1_a : a \to a$;
- The operation of *composition*, which assigns to each pair $\langle g, f \rangle$ of arrows with $\operatorname{dom} g = \operatorname{cod} f$ an arrow $g \circ f$ called their *composite*. Note that $g \circ f : \operatorname{dom} f \to \operatorname{cod} g$ (see Figure 11.2).

11.2.2 Operations in a Category

The operations in a category satisfy two axioms:

Axiom 1 (Associativity) For given objects and arrows such that

$$a \xrightarrow{f} b \xrightarrow{g} c \xrightarrow{k} d,$$

one always has the equality

$$k \circ (g \circ f) = (k \circ g) \circ f.$$

This relationship can also be represented by a diagram (Figure 11.3):

Axiom 2 (Unit Law) For all arrows $f : a \to b$ and $g : b \to c$, composition with the identity arrow $\mathbf{1}_b$ gives

$$\mathbf{1}_b \circ f = f \qquad \text{and} \qquad g \circ \mathbf{1}_b = g.$$

Thus the identity arrow $\mathbf{1}_b$ of each object b acts as an identity for the operation of composition whenever that composition makes sense (see Figure 11.4).

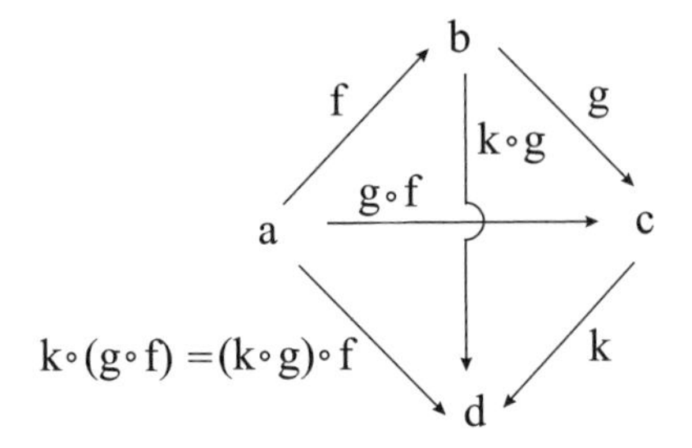

Figure 11.3

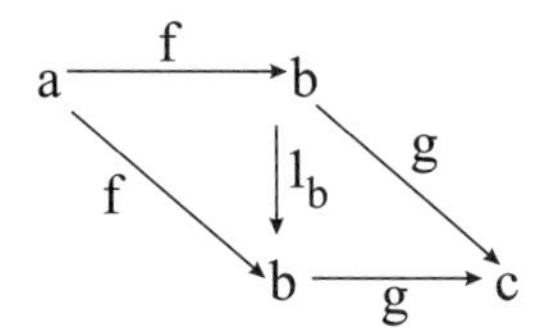

Figure 11.4

11.2.3 Commutative Diagrams

Category theory uses many diagrams such as the four that we have seen thus far. Such a diagram is said to be *commutative* if, for each pair of vertices c and c', any two paths formed from directed edges leading from c to c' yield, by composition of labels, equal arrows from c to c'.

11.2.4 Arrows Instead of Objects

Since an object b in a metacategory corresponds exactly to its identity map $\mathbf{1}_b$, it is possible to dispense altogether with the objects and to deal only with the arrows. Such a rendition C of metacategory theory consists of arrows, certain ordered pairs $\langle g, f\rangle$ (which we think of as the "composable pairs" of arrows), and an operation assigning to each composable pair $\langle g, f\rangle$ an arrow $g \circ f$ which we call their composite. One then defines an identity of C to be an arrow u such that $f \circ u = f$ whenever the composite $f \circ u$ is defined and $u \circ g = g$ whenever $u \circ g$ is defined. The data are then required to satisfy these axioms:

Axiom (a) The composite $(k \circ g) \circ f$ is defined if and only if the composite $k \circ (g \circ f)$ is defined; when either is defined, then they are equal. We typically write the composite as $k \circ g \circ f$.

Axiom (b) The triple composite $k \circ g \circ f$ is defined whenever both composites $k \circ g$ and $g \circ f$ are defined.

Axiom (c) For each arrow g of C, there are identity arrows u and u' of C such that $u' \circ g$ and $g \circ u$ are defined.

11.2.5 Metacategories and Morphisms

A metacategory is to be any interpretation that satisfies the three axioms specified above. An example would be the metacategory of sets, which has as objects all sets and as arrows all functions, with the usual notions of composition and identity. Other examples would be the metacategory of groups and the metacategory of topological spaces.

It is common to refer to the arrows in a metacategory as its *morphisms*.

11.2.6 Categories

A *category* is any interpretation of the category axioms within set theory. This idea requires some explanation.

A *graph* or *diagram scheme* is a set O of objects, a set A of arrows, and two functions

$$A \stackrel{\text{dom}}{\to} O$$

and

$$O \stackrel{\text{cod}}{\to} A.$$

In the graph, the set of composable pairs of arrows is the set

$$A \times_O A \equiv \{\langle g, f\rangle : g, f \in A \text{ and } \operatorname{dom} g = \operatorname{cod} f\}.$$

This set is called the *product over* O.

11.2.7 Categories and Graphs

A *category* is a graph with two additional functions:

$$\begin{aligned} O &\stackrel{\text{id}}{\to} A \\ c &\longmapsto \text{id}_c \end{aligned}$$

and

$$\begin{aligned} A \times_O A &\stackrel{\circ}{\to} A \\ \langle g, f\rangle &\mapsto g \circ f. \end{aligned}$$

These are called, respectively, the identity and composition functions. We require that

$$\operatorname{dom}(\text{id}\, a) = a = \operatorname{cod}(\text{id}\, a),$$

$$\operatorname{dom}(g \circ f) = \operatorname{dom} f,$$

$$\operatorname{cod}(g \circ f) = \operatorname{cod} g,$$

for all objects a in the graph O and for all composable pairs of arrows $\langle g, f\rangle \in A \times_O A$. And we require that the Associativity Axiom and Unit Axiom of Subsection 11.2.2 hold.

In practice, we do not refer explicitly to A and O but instead say (for brevity) that $c \in C$ and $f \in C$. We also write

$$\hom(b, c) = \{f : f \in C, \operatorname{dom} f = b, \operatorname{cod} f = c\}$$

to denote the set of arrows from b to c.

11.2.8 Elementary Examples of Categories

Example 11.1

The symbol $\mathbf{0}$ denotes the empty category, with no objects and no arrows.

□

Example 11.2

The symbol $\mathbf{1}$ denotes the category with one object and one arrow (the identity arrow).

□

Example 11.3

The symbol $\mathbf{2}$ denotes the category with two objects, a and b, and just one arrow $a \to b$, which is not the identity.

□

Example 11.4

The symbol $\mathbf{3}$ denotes the category with three objects whose nonidentity arrows are arranged as in the triangle depicted in Figure 11.5.

□

Example 11.5

The symbol $\downarrow\downarrow$ denotes the category with two objects a and b and just two arrows $a \to b$ and $b \to a$, which are not the identity arrows. We call two such arrows "parallel arrows."

□

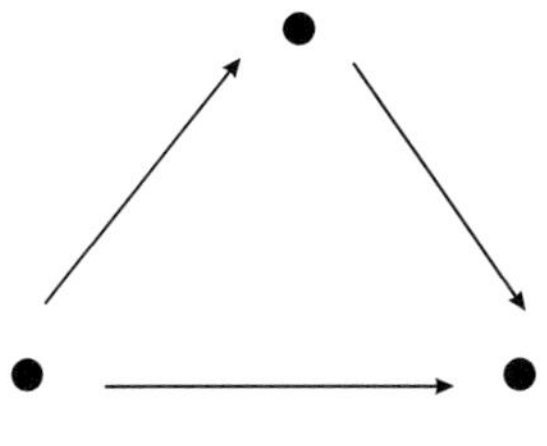

Figure 11.5

11.2.9 Discrete Examples of Categories

A category is *discrete* if every arrow is an identity. Every set X is the set of objects of a discrete category, and every discrete category is determined by its set of objects.

Example 11.6

(Monoids) A *monoid* is a category with just one object. Each monoid is determined by the set of all of its arrows, by the identity arrow, and by the rule for the composition of arrows. Since any two arrows therefore have a composition, we may think of the monoid as a set M with a binary operation $M \times M \to M$, that is associative and has an identity (i.e., a unit) element. In other words, a monoid is a semigroup with identity. For any category C and any object $a \in C$, the set $\hom(a, a)$ of all arrows $a \to a$ is a monoid.

□

Example 11.7

(Groups) A *group* is a category with one object (i.e., the set of elements of the group) in which every arrow has a two-sided inverse under composition.

□

Example 11.8

(Matrices) Let K be a commutative ring. Then the set $\mathbf{Matr}_K$ of all rectangular matrices (of any dimensions) with entries in K is a category. All of the objects are positive integers $m, n, \ldots$. Each $m \times n$ matrix A is treated as an arrow $A : n \to m$, with composition given by the ordinary matrix product.

□

Example 11.9

(Sets) If V is any collection of sets, then we take Ens_V to be the category with all sets $X \in V$ as the objects. We take the arrows to be all functions $f : X \to Y$, with the usual notion of composition.

□

11.2.10 Functors

A *functor* is a morphism (i.e., an arrow) of categories. More precisely, if C and B are categories, then a functor $T : C \to B$ with domain C and codomain B consists of two related functions. These are

(i) the *object function* T, which assigns to each object $c \in C$ an object $T(c)$ of B;

(ii) the *arrow function*, which assigns to each arrow $f : c \to c'$ of C an arrow $Tf : Tc \to Tc'$ of B so that

$$T(\mathbf{1}_c) = \mathbf{1}_{Tc}$$
$$T(g \circ f) = Tg \circ Tf \text{ whenever } g \circ f \text{ is defined in } C.$$

We also denote the arrow function by T.

Example 11.10

Consider the *power set functor* $\mathcal{P} : \mathbf{Set} \to \mathbf{Set}$, which assigns to each set X its usual power set. The arrow function assigns to each $f : X \to Y$ that map $\mathcal{P}f : \mathcal{P}X \to \mathcal{P}Y$ that sends each $S \subseteq X$ to its image $fS \subseteq Y$. Clearly, $\mathcal{P}(\mathbf{1}_X) = \mathbf{1}_{\mathcal{P}X}$ and $\mathcal{P}(g \circ f) = \mathcal{P}g \circ \mathcal{P}f$.

□

Example 11.11

Fix a dimension n. Assign to each topological space X an abelian group $H_n(X)$, the nth singular homology group of X. Then to each continuous map $f : X \to Y$ of topological spaces there corresponds a homomorphism $H_n(f) : H_n(X) \to H_n(Y)$ of groups in such a way that the operator H_n becomes a functor $\mathbf{Top} \to \mathbf{Ab}$.

□

11.2.11 Natural Transformations

Let S, T be functors from category C to category B. Then a *natural transformation* $\tau : S \dot{\to} T$ is a function that assigns to each object $c \in C$

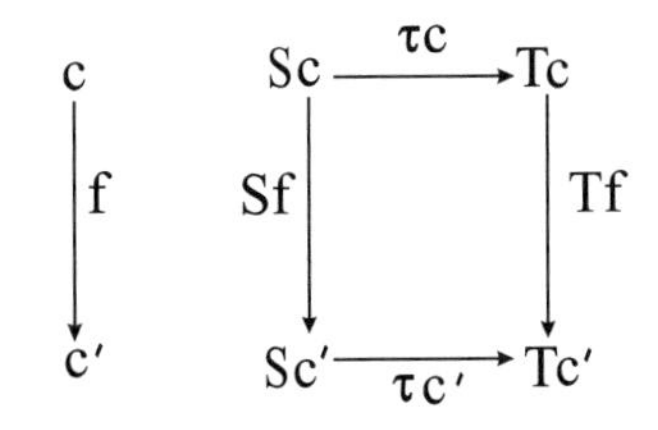

Figure 11.6

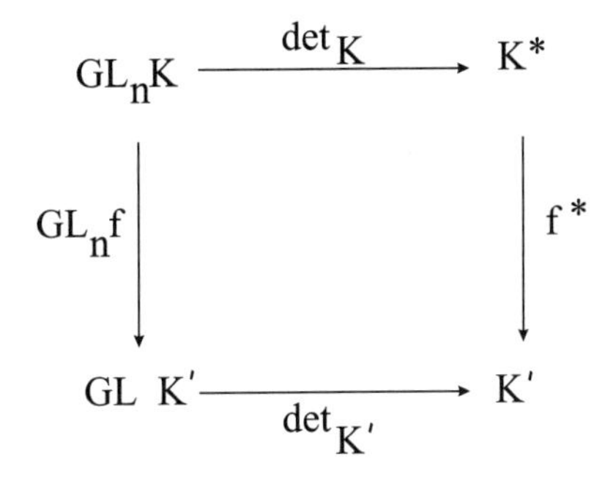

Figure 11.7

an arrow $\tau_c \equiv \tau c : S(c) \to T(c)$ of B so that every arrow $f : c \to c'$ in C yields a commutative diagram (Figure 11.6). Under these circumstances, we will also say that $\tau_c : S(c) \to T(c)$ is *natural* in c.

Example 11.12

The ordinary determinant from elementary linear algebra is a natural transformation. For let $\det_K M$ be the determinant of the $n \times n$ matrix M with entries from the commutative ring K, and let K^* denote the group of units of K. So M is nonsingular when $\det_K M$ is a unit, and $\det_K$ is a group morphism from $GL_n K$ to K^*. *Because the determinant is defined by the same formula for all rings* K, each morphism $f : K \to K'$ of commutative rings leads to a commutative diagram (Figure 11.7). In summary, the transformation $\det : GL_n \to (\ )^*$ is natural between two functors **CRng** $\to$ **Grp**.

□

Example 11.13

Let G be a group. Then the projection $p_G : G \to G/[G, G]$ from G to the factor-commutator group defines a transformation from the identity functor on **Grp** to the factor-commutator functor **Grp** $\to$ **Ab** $\to$ **Grp**. Note that p is natural because each group

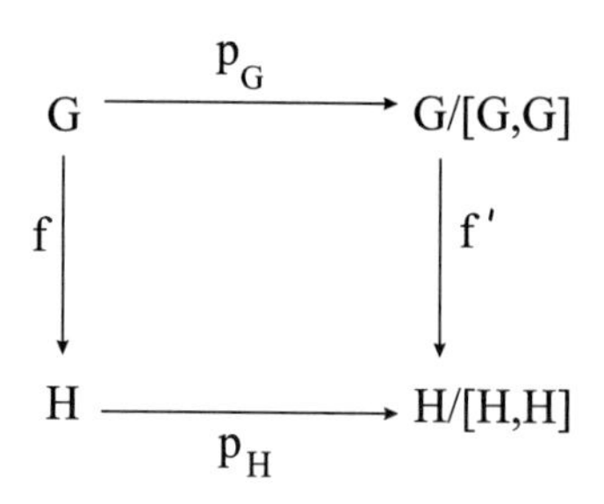

Figure 11.8

homomorphism $f : G \to H$ defines the obvious homomorphism f; so that the diagram shown in Figure 11.8 commutes.

□

11.2.12 Algebraic Theories

The use of the categorical presentation of algebraic theory leads to a systematic way to interpret one equational theory into another, as well as to a theory of categories of algebras. We draw on the exposition in [BAR, pp. 283–313].

Definition 11.1 An *algebraic theory* T is a category whose objects are the natural numbers $\mathbb{N}$ and which for each $n \in \mathbb{N}$ is equipped with an n-tuple of maps

$$\text{proj} : n \to 1, \quad i = 1, \ldots, n;$$

this makes n into the n-fold categorical product of 1, $n = 1^n$.

Example 11.14

Define an algebraic theory T by

$$\text{hom}(n, m) = m\text{-tuples of polynomials in } X_1, \ldots, X_n \text{ with integral coefficients.}$$

We use substitution of polynomials as the composition. Note that the "proj_i" is just X_i, considered as a polynomial in the variables $X_1, \ldots, X_n$, for $1 \leq i \leq n$. We call this T "the theory of commutative rings."

□

In any algebraic theory, we can describe a map $n \to m$ by an m-tuple of maps $n \to 1$, just because m is an m-fold product of 1. Thus the maps

$n \to 1$ play a special role; they are the n-ary operations in the theory. In the example, an n-ary operation is a polynomial in n variables.

Definition 11.2 Suppose that T is an algebraic theory. Let $\mathcal{E}$ be a category with finite products. The *category of T-algebras* or *T-models* in $\mathcal{E}$ is the entire subcategory $\mathbf{Alg}(T,\ \mathcal{E})$ of the functor category $(T, \mathcal{E})$; its objects are the finite, product-preserving functions.

Definition 11.3 A *morphism* of algebraic theories is a functor $f : T \to T'$ such that $f(n) = n$ and $f(\mathrm{proj}_i) = \mathrm{proj}'_i$ for all of the given projections.

Chapter 12

Complexity Theory

Man propounds negotiations, man accepts the compromise. Very rarely will he squarely push the logic of a fact to its ultimate conclusion in unmitigated act.

—Rudyard Kipling

No man, for any considerable period, can wear one face to himself, and another to the multitude, without finally getting bewildered as to which may be the true.

—Nathaniel Hawthorne

Light; or, failing that, lightning: the world can take its choice.

—Thomas Carlyle

Things and men have always a certain sense, a certain side by which they must be got hold of if one wants to obtain a solid grasp and a perfect command.

—Joseph Conrad

In what we really understand, we reason but little.

—William Hazlitt

It takes a long time to understand nothing.

—Edward Dahlberg

All men naturally desire knowledge.

—Aristotle

In all questions of logical analysis, our chief debt is to Frege.

—Bertrand Russell and Alfred North Whitehead

12.1 Preliminary Remarks

Complexity theory is a means of measuring how complicated it is, or how much computer time it will take, to solve a problem. We measure complexity theory in the following way: Suppose that the formulation of

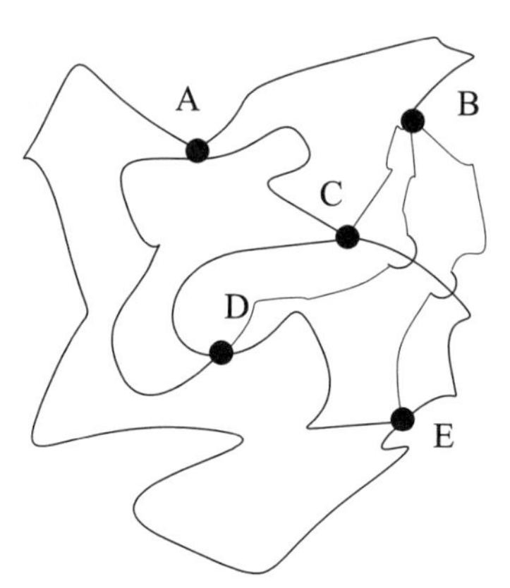

Figure 12.1

an instance of a problem involves n pieces of data. Then how many steps will it take (as a function of n) to solve the problem? Can we obtain an effective bound on that number of steps that is valid for asymptotically large values of n?

12.2 Polynomial Complexity

Consider dropping n playing cards on the floor. Your job is to put them back in order. How many steps will this take?

In at most n steps (just by examining each card), you can locate the first card. In at most another $(n-1)$ steps, you can locate the second card. In at most another $(n-2)$ steps, you can locate the third card, and so forth. In summary, it will require at most

$$n + (n-1) + (n-2) + \cdots + 1 = \frac{n(n+1)}{2}$$

steps to put the deck of cards back in order (this summation formula is proved in Subsection 8.2.1). Because the expression $n[n+1]/2 \leq 2n^2$, we say that this problem has *polynomial complexity of degree (at most) 2.*

12.3 Exponential Complexity

Now, contrast that first example with the celebrated traveling salesman problem: There are given n cities and there is a road of known length connecting each pair (see Figure 12.1). The problem is to find the shortest route that will enable the salesman to visit each city and to return to his starting place.

On the level of effective computability (see Subsections 6.2.1 and 6.2.2), we see that a search for the solution of the traveling salesman problem amounts to examining each possible ordering of cities (because clearly the salesman can visit the cities in any order). There are $n!$ such orderings. According to Stirling's formula,

$$n! \approx \left(\frac{n}{e}\right)^n \cdot \sqrt{2\pi n}.$$

Thus the problem cannot be solved in a polynomial number of steps. We say that the problem has *exponential complexity.*

Another famous problem that is exponentially complex is the *subgraph problem.* Let G be a graph; that is, G is a collection of vertices together with certain edges that connect certain pairs of the vertices. Let H be another graph. The question is whether G contains a sub-graph that is isomorphic to H. (Here two graphs are isomorphic if there is a combinatorial mapping matching up vertices and edges.)

It may be noted that there are orders of magnitude that lie strictly between polynomial size and exponential size. For example, $n^{\log n}$ is (asymptotically) strictly greater than any polynomial, but it is (asymptotically) strictly less than any exponential function. In practice, in theoretical computer science, we tend to refer to any algorithm that is not of polynomial size as an exponential growth algorithm—even though this terminology is not strictly correct.

12.4 Two Tables for Complexity Theory

12.4.1 Table Illustrating the Difference Between Polynomial and Exponential Complexity

Table 12.1 illustrates, for various large values of n, the difference between some polynomial degrees of complexity and some degrees of exponential complexity. Times given are on a theoretical computing machine. The data are taken from [GAJ, p. 7].

12.4.2 Problems That Can Be Solved in One Hour

Table 12.2, also an idea from [GAJ, p. 8], illustrates the largest problem that can be solved in one hour—with the theoretical machine used in the first column, with a machine 100 times faster in the second column, and with a machine 1000 times faster in the last column.

Observe that, with exponential complexity, the increase in the size of a problem that can be solved in one hour using a machine that is even 1000 times faster is only marginal. That is because the increase is in fact the logarithm of the speed factor.

time cplxty. fcn.	$n = 10$	$n = 20$	$n = 30$	$n = 40$	$n = 50$	$n = 60$
n	.00001 second	.00002 second	.00003 second	.00004 second	.00005 second	.00006 second
n^2	.0001 second	.0004 second	.0009 second	.0016 second	.0025 second	.0036 second
n^3	.001 second	.008 second	.027 second	.064 second	.125 second	.216 second
n^5	.1 second	3.2 seconds	24.3 seconds	1.7 minutes	5.2 minutes	13.0 minutes
2^n	.001 second	1.0 second	17.9 minutes	12.7 days	13030.5 days	366 centuries
3^n	.059 second	58 minutes	2372.5 days	3855 centuries	2×10^8 centuries	1.3×10^{13} centuries

Table 12.1

time complexity function	with present computer	with computer 100 times faster	with computer 1000 times faster
n	N_1	$100N_1$	$1000N_1$
n^2	N_2	$10N_2$	$31.6N_2$
n^3	N_3	$4.64N_3$	$10N_3$
n^5	N_5	$2.5N_5$	$3.98N_5$
2^n	N_{exp}	$N_{\text{exp}} + 6.64$	$N_{\text{exp}} + 9.97$
3^n	$N_{\text{exp}'}$	$N_{\text{exp}'} + 4.19$	$N_{\text{exp}'} + 6.29$

Table 12.2: Size of Largest Problem Solvable in One Hour

12.4.3 Comparing Polynomial and Exponential Complexity

From the point of view of theoretical computer science, it is a matter of considerable interest to know whether a problem is of polynomial or of exponential complexity, for this information gives an indication of how computationally expensive a certain procedure will be. In the area of computer games (for example), this will translate into whether a certain action can be executed with realistic speed, or sluggishly.

In the discussion that follows, we restrict attention to the so-called "decision problems." These are problems with yes/no answers. An example of a problem that is *not* a decision problem is an optimization problem (e.g., find the configuration that maximizes something). But in fact most optimization problems can be converted to decision problems by introducing an auxiliary parameter that plays the role of an upper bound. It is not a severe restriction to treat only decision problems, and it makes the exposition much cleaner.

12.5 Problems of Class **P**

12.5.1 Polynomial Complexity

A problem is said to be of "Class **P**" if there is a polynomial p and a (deterministic) Turing machine (DTM) for which every input of length n comes to a halt, with a yes/no answer, after at most $p(n)$ steps. The word "deterministic" is used here to denote an effectively computable process, with no guessing. The Turing machine that we described in Subsection 6.3.2 is deterministic.

12.5.2 Tractable Problems

Problems of Class **P** are considered to be tractable. A problem that is not of class **P**—that is, for which there is no polynomial solution algorithm—is by definition *intractable*. As the tables in Subsections 12.4.1 and 12.4.2 indicate, Class **P** problems give solutions in a reasonable amount of time, and *they always give solutions*. It will never happen that the machine runs forever. The problem of ordering a deck of cards (already discussed), the problem of finding one particular marble in a jar of marbles, and the problem of matching up husbands and wives at a party are all problems of Class **P**.

12.5.3 Problems That Can Be Verified in Polynomial Time

More significant for the theoretical development of this chapter is that, for certain problems that are otherwise intractable, the *verification of a solution* can be a problem of Class **P**. We describe the details of this assertion below.

For example, the problem of finding the prime factorization of a given natural number N with n digits is known to be of exponential complexity (see [SCL]). More precisely, the complexity is about of size $10^{\sqrt{n \ln n}}$, which means that the computation would take about that many steps. But the verification procedure is of polynomial complexity: If N is given and a putative factorization $p_1, \ldots, p_k$ is given, then it is obviously (just by inspection of the rules of arithmetic) a polynomial time problem to calculate $p_1 \cdot p_2 \cdots p_k$ and verify (or not) that it equals N.

Likewise, the subgraph problem is known to be of exponential complexity. But, if one is given a graph G, a subgraph H, and another graph K, then it is a problem of only polynomial complexity to confirm (or not) that H is graph-theoretically isomorphic to K.

The considerations in the last three paragraphs will play a decisive role in our development of the concept of problems of "Class **NP**."

12.6 Problems of Class NP

12.6.1 Nondeterministic Turing Machines

A nondeterministic Turing machine (NDTM) is a Turing machine with an extra write-only head that generates an initial guess for the Turing machine to evaluate (see [GAJ] for a rigorous definition of the concept of nondeterministic Turing machine). We say that a problem is of Class **NP** if there is a polynomial p and a nondeterministic Turing machine with the property that, for any instance of the problem, the Turing machine will generate a guess of some length n and come to a halt, generating an answer of "yes" or "no" (i.e., that this guess *is* a solution or *is not* a solution) in at most $p(n)$ steps.

12.6.2 NP *Contains* P

It first should be noted that $\mathbf{P} \subseteq \mathbf{NP}$. This assertion is obvious, because if Π is a problem of class **P**, then we can let the guess be vacuous to see that Π is then of class **NP**. It is natural, therefore, to consider $\mathbf{NP} \setminus \mathbf{P}$.

The first interesting question in this subject is whether $\mathbf{NP} \setminus \mathbf{P}$ is nonempty. If it is, then any problem in that set-theoretic difference has the property that it is *not* deterministically of polynomial complexity, but if it is given a guess for a solution, then it can evaluate that guess in polynomial time. Any problem that can be established to lie in this set-theoretic difference will be considered to be intractable.

12.6.3 The Difference Between **NP** *and* **P**

It turns out that it is quite difficult to determine whether $\mathbf{NP} \setminus \mathbf{P}$ is nonempty. Thus far, no such problem has been identified. So some alternative questions have been formulated; these seem to be more within our reach. In particular, there are relatively straightforward proof techniques for addressing these alternative formulations.

12.6.4 Foundations of **NP**-*Completeness*

The first of these separate questions, which is the foundational question of **NP**-completeness, is the following. Suppose that we are given a problem Π. Can we establish the following syllogism?

If $\mathbf{NP} \setminus \mathbf{P}$ is nonempty (i.e., $\mathbf{P} \neq \mathbf{NP}$), then $\Pi \in \mathbf{NP} \setminus \mathbf{P}$.

See [GAJ] for sample problems that may be addressed using this slightly modified problem.

The most important substitute question is that of **NP**-completeness. We will address it in the next section.

12.6.5 Limits of the Intractability of **NP** *Problems*

We close the present section by recording a remarkable result about **NP** problems. It says, in effect, that an **NP** problem may be intractable, but there is a limit to its intractability.

Theorem 12.1
If $\Pi \in \mathbf{NP}$*, then there is a polynomial* p *such that* Π *has a deterministic solution algorithm having complexity no greater than* $C \cdot 2^{p(n)}$*. That is, if input data to the problem have length* n*, then the solution procedure will have length not greater than a constant* C *times* $2^{p(n)}$*.*

The theory of **NP** problems connects with the theory of finite models (Section 3.2.3) by way of a famous result of Fagin: the class **NP** is just the same as the class of generalized spectra. Here the spectrum of a first-order sentence is the set of cardinalities of its finite models. See [STE] for more on this matter.

12.7 **NP**-*Completeness*

12.7.1 Polynomial Equivalence

We say that two problems Π_1 and Π_2 are *polynomially equivalent* if there is a polynomially complex translation of the language in which

Π_1 is expressed into the language in which Π_2 is expressed. That is to say, there is a polynomial q such that any statement about Π_1 with n characters can be translated into a statement about Π_2 with at most $q(n)$ characters and conversely.

It is worth noting that any two of the standard models of computation are polynomially equivalent.

12.7.2 Definition of **NP**-Completeness

Let Π be a problem in **NP**. We say that Π is **NP**-*Complete* if Π is polynomially equivalent to every other problem in **NP**. Now suppose that Π is an **NP**-complete problem. If it turns out that Π can be solved with a polynomial time algorithm, then it follows that every other problem in **NP** can be solved with a polynomial time algorithm. On the other hand, if it turns out that *any* problem in **NP** is intractable (i.e., lies in $\mathbf{NP} \setminus \mathbf{P}$), then Π is intractable. The **NP**-complete problems are considered to be the hardest problems in **NP**.

12.7.3 Intractable Problems and **NP**-Complete Problems

Although we do not know whether $\mathbf{NP} \setminus \mathbf{P}$ is nonempty (i.e., whether there are any intractable problems), we do know—and can explicitly identify—a great many **NP**-complete problems. It is considered to be one of the great unsolved problems in theoretical computer science and mathematics to determine whether **NP**-complete problems are intractable. Thus far, little substantial progress has been made on the question.

12.7.4 Structure of the Class **NP**

It is *not* the case that **NP** partitions into problems of class **P** and problems that are **NP**-complete. In fact, it can be proved that if $\mathbf{P} \neq \mathbf{NP}$, then there will exist problems that are both *not* of class **P** and *not* **NP**-complete.

In Section 12.9 we will give brief descriptions of several problems that are known to be **NP**-complete. In fact, the monograph [GAJ] lists over 300 such problems, and many more have been discovered since the publication of that text.

12.7.5 The Classes Pspace and Log-Space

There is a new complexity class called *pspace*. It consists of those languages that can be accepted by a Turing machine in polynomial space. It is known that **NP** problems lie in pspace. It is not known whether that inclusion is proper.

Another interesting class is the problems of log-space. These are problems with n pieces of input that require space of size on the order of $\log n + 1$ for solution. It is known that the deterministic log-space problems are of class **P**. It is not known whether that inclusion is proper (see [GAJ, p. 177]).

12.8 Cook's Theorem

Stephen Cook laid the foundations for the theory of **NP**-completeness in his paper [COO]. One of the seminal results of that work is Cook's theorem, which we now describe.

12.8.1 The Satisfiability Problem

Let $\mathcal{P}$ be a finite collection of statements from the propositional calculus. That is, it is a collection of sentences formed from finitely many atomic sentences $A_1, \ldots, A_k$ using the basic connectives. The satisfiability problem is this: Is there an assignment of truth values to $A_1, \ldots,$ A_k so that all statements in $\mathcal{P}$ are true?

To give two simple examples:

- The collection of statements $\mathcal{P} = \{A \vee \sim B, \sim A \vee B\}$ is satisfiable. Indeed, the truth assignment $t(A) = T$, $t(B) = T$ will do the job.

- The collection of statements $\mathcal{P} = \{A \vee B, A \vee \sim B, \sim A\}$ is not satisfiable. For A must be true in order for the first two statements to be true, and then $\sim A$ is false.

12.8.2 Enunciation of Cook's Theorem

Now Cook's theorem grants the satisfiability problem a seminal role in the theory of **NP**-completeness. In fact, it was historically the first problem to be determined to be **NP**-complete.

Theorem 12.2 (Cook)
The satisfiability problem is **NP***-complete.*

12.9 Examples of **NP**-Complete Problems

Here we enumerate several examples, from many different branches of mathematics and theoretical computer science, of problems that are known to be **NP**-complete. Bear in mind that if any of these problems is determined to be intractable, then they all are. Our source for this material is the Appendix of [GAJ].

12.9.1 Problems from Graph Theory

In what follows, if G is a graph, we write $G = (V, E)$, where V denotes the collection of vertices and E denotes the collection of edges of the graph. The notation $|V|$ denotes the number of vertices in the graph. A graph is said to be *connected* if it is not the disjoint union of two subgraphs.

Complete Subgraphs: The *complete graph on k vertices* is a graph with k vertices such that *every* pair of vertices is connected by an edge. Thus the complete graph on k vertices has $\binom{k}{2} = \frac{k(k-1)}{2}$ edges.

> **Q**: Given a graph $G = (V, E)$ and a natural number $k \leq |V|$, does G contain a subgraph that is isomorphic to the complete graph on k vertices?

Reference: [COO].

Graph Coloring Problem: Let $G = (V, E)$ be a graph. A *coloring* of G is an assignment of finitely many colors, one to each vertex of G, so that if two vertices are connected then they have different colors.

> **Q**: Given a graph $G = (V, E)$ and a natural number $3 \leq k \leq |V|$, can G be colored with k colors?

Reference: [KAR].

Monochromatic Triangle Problem: If $G = (V, E)$ is a graph, then we say that G contains a *triangle* if there are three vertices e_1, e_2, e_3 such that any two of these vertices are connected by an edge.

> **Q**: Given a graph $G = (V, E)$, is there a partition of E into two sets, $E = E_1 \cup E_2$, so that neither $G_1 = (V, E_1)$ nor $G_2 = (V, E_2)$ contains a triangle?

Reference: [BUR].

Partition into Triangles Problem: Let $G = (V, E)$ be a graph. Let the term "triangle" be as in the Monochromatic Triangle Problem.

> **Q**: Let $G = (V, E)$ be a graph. Can V be partitioned into disjoint sets, $V = V_1 \cup \cdots V_k$, each containing exactly three vertices, so that each V_j contains exactly three vertices and these vertices form a triangle in G?

Reference: [SCHA1].

The Problem of Cliques: Let $G = (V, E)$ be a graph and let k be a natural number with $k \leq |V|$. A *clique* of size k in G is a subset $V' \subseteq V$ with $|V'| = k$ such that every two vertices in V' are joined by an edge in E.

> **Q**: Does the graph G contain a clique of size k or more?

Reference: [KAR].

The Independent Set Problem: Let $G = (V, E)$ be a graph and let k be a natural number with $k \leq |V|$. An *independent set* of size k is a subset $V' \subseteq V$ such that $|V'| = k$ and no two vertices in V' are joined by an edge in E.

> **Q**: Does G contain an independent set of size k or more?

References: [KAR] and [HOK].

Planar Subgraph Problem: Let $G = (V, E)$. We say that G is *planar* if there is a set of points V' in the plane and a collection of edges E' in the plane so that the graph $G' = (V', E')$ is combinatorially equivalent to G. (It should be understood here that the edges in E' in the plane are not allowed to cross. As an example, the complete graph on five vertices is not planar.)

> **Q**: Let $G = (V, E)$ be a graph and k a natural number. Is there a subset $E' \subseteq E$ with $|E'| \geq k$ such that $G' = (V, E')$ is planar?

Reference: [LIG].

12.9.2 Problems from Network Design

Degree Constrained Spanning Tree Problem: Let $G = (V, E)$ be a graph. A *spanning tree* for G is a subgraph H that includes all of the vertices, is connected, and has no circuits (i.e., there is only one path connecting any two vertices).

If $x \in V$ is a vertex of G, then the *degree* of x is the number of ends of edges that occur at x.

> **Q**: Let $G = (V, E)$ be a graph and k a natural number with $k \leq |V|$. Is there a spanning tree for G in which no vertex has degree larger than k?

References: [KAR] and [LIG].

Maximum Leaf Spanning Tree Problem: The terms "spanning tree" and "degree" were defined in the last problem.

> **Q**: Let $G = (V, E)$ be a graph and k a natural number with $k \leq |V|$. Is there a spanning tree for G in which k or more vertices have degree 1?

Reference: [GAJ, p. 75].

Isomorphic Spanning Tree Problem: A *tree* is a connected graph that contains no cycles.

> **Q**: Let $G = (V, T)$ be a graph and $T = (V_T, E_T)$ be a tree. Does G contain a spanning tree that is isomorphic to T?

References: [KAR] and [LIG].

12.9.3 Problems from the Theory of Sets and Partitions

Set Packing Problem:

> **Q**: Let $\mathcal{C}$ be a collection of finite sets and k a natural number with $k \leq |C|$. Does $\mathcal{C}$ contain at least k mutually disjoint sets?

Reference: [KAR].

Set Splitting Problem:

> **Q**: Let $\mathcal{C}$ be a collection of subsets of the finite set S (in other words, $\mathcal{C}$ is a subset of the power set of S). Is there a partition $S = S_1 \cup S_2$, with $S_1 \neq \emptyset, S_2 \neq \emptyset$, such that each element of $\mathcal{C}$ is entirely contained in either S_1 or S_2?

Reference: [LOV].

Problem of the kth Largest Subset:

> **Q**: Let a finite set A and two natural numbers k and B be given. Also assume that there is a function $s : A \to \mathbb{N}$ that assigns a "size" to each element of A. Are there ℓ distinct subsets $S_1, S_2, \ldots, S_\ell$ of A, with $\ell \geq k$, such that the sum of the sizes of the elements of each S_j does not exceed B?

Reference: [JOK]. This problem is not even known to be in **NP**. We refer to such a problem as "**NP**-hard."

Hitting Set Problem:

> **Q**: Let $\mathcal{C}$ be a collection of subsets of a given finite set S. Let k be a natural number with $k \leq |S|$. Is there a subset $S' \subseteq S$ with $|S'| \leq k$ such that S' contains at least one element from each element of $\mathcal{C}$?

Reference: [KAR].

Intersection Pattern Problem:

> **Q**: Let F be an $n \times n$ matrix with entries f_{ij} in $\mathbb{N}$. Is there a collection $\{S_1, \ldots, S_n\}$ of sets such that, for all $i, j = 1, \ldots, n$, $f_{ij} = |S_i \cap S_j|$?

Reference: [KAR].

12.9.4 Storage and Retrieval Problems

Bin Packing Problem:

> **Q**: Let U be a finite set of items. Let $s : U \to \mathbb{N}$ be a function that assigns a size to each element of U. Let $B, k \in \mathbb{N}$. Is there a partition $U = U_1 \cup \cdots \cup U_k$ such that $\sum_{u \in U_j} s(u) \leq B$ for each j?

References: [DAN], [HOS], and [LAW].

Sparse Matrix Compression Problem:

> **Q:** Let F be an $m \times n$ matrix with entries f_{ij} in $\{0, 1\}$ and let $3 \le k \le mn$ be an integer. Is there a sequence $\{b_1, b_2, \ldots, b_{n+k}\}$ of integers, each satisfying $0 \le b_i \le m$, and a function $s : \{1, 2, \ldots, m\} \to \{1, 2, \ldots, k\}$ such that, for each $1 \le i \le m$ and $1 \le j \le n$, the entry $f_{ij} = 1$ if and only if $b_{s(i)+j-1} = i$?

Reference: [ELS].

Consecutive Ones Matrix Partition Problem:

We say that a matrix A of 0's and 1's has the *consecutive ones property* if the columns of A can be permuted so that, in each row, all the 1's occur consecutively.

> **Q:** Let F be an $m \times n$ matrix of 0's and 1's. Can the rows of F be partitioned into two groups such that the resulting $m_1 \times n$ and $m_2 \times n$ matrices ($m_1 + m_2 = m$) each have the consecutive ones property?

Reference: [LIP].

Rectilinear Picture Compression Problems:

> **Q:** Let $F = \{f_{ij}\}$ be an $n \times n$ matrix of 0's and 1's and let $k \in \mathbb{N}$. Is there a collection of k or fewer rectangles that covers precisely those entries in F that are 1's in the following sense: Is there a sequence of quadruples (a_i, b_i, c_i, d_i), $1 \le i \le k$, with $a_i \le b_i$ and $c_i \le d_i$ for $1 \le i \le k$, such that for each pair (i, j) we have the property $f_{ij} = 1$ if and only if there exists an index ℓ, $1 \le \ell \le k$, such that $a_\ell \le i \le b_\ell$ and $c_\ell \le j \le d_\ell$?

Reference: [MAK].

12.9.5 Sequencing and Scheduling Problems

Staff Scheduling Problem:

> **Q:** Let $1 \le k \le m \in \mathbb{N}$. Let C be a collection of m-tuples, each having k entries 1 and $m - k$ entries 0. Let R be an m-tuple of nonnegative integers, and let $n \in \mathbb{N}$. Is there a function $f : C \to \mathbb{Z}^+$ such that $\sum_{c \in C} f(c) \le n$ and $\sum_{c \in C} f(c) \cdot c \ge R$?

Reference: [GAJ, p. 243].

Production Planning Problem:

> **Q**: Let $n \in \mathbb{N}$ and, for each $1 \leq i \leq n$, numbers $r_i \in \mathbb{Z}^+$, $c_i \in \mathbb{Z}^+$, $b_i \in \mathbb{Z}^+$, $p_i \in \mathbb{Z}^+$, and $h_i \in \mathbb{Z}^+$. Let $B \in \mathbb{N}$. Do there exist numbers $x_i \in \mathbb{Z}^+$ and $I_i = \sum_{j=1}^{i}(x_j - r_j)$, $1 \leq i \leq n$ such that all $x_i \leq c_i$, all $I_i \geq 0$, and
> $$\sum_{i=1}^{n}(p_i x_i + h_i I_i) + \sum_{x_i > 0} b_i \leq B?$$

Reference: [LRKF].

12.9.6 Problems from Mathematical Programming

Integer Programming Problem:

> **Q**: Let X be a finite set of pairs (x, b), where x is an m-tuple of integers and b is an integer. Let c be an m-tuple of integers and B another integer. Is there an m-tuple y of integers such that $x \cdot y \leq b$ for all $(x, b) \in X$ and such that $c \cdot y \geq B$?

References: [KAR] and [BOT].

Knapsack Problem:

> **Q**: Let U be a finite set, $s : U \to \mathbb{N}$, and $v : U \to \mathbb{N}$. Let B, k be positive integers. Is there a subset $U' \subseteq U$ such that $\sum_{u \in U'} s(u) \leq B$ and $\sum_{u \in U'} v(u) \geq k$?

Reference: [KAR].

Open Hemisphere Problem:

> **Q**: Let X be a finite set of m-tuples of integers. Let $k \in \mathbb{N}$, $k \leq |X|$. Is there an m-tuple y of rational numbers such that $x \cdot y > 0$ for at least k of the m-tuples $x \in X$?

Reference: [JOP].

Comparative Vector Inequality Problem: If u and v are m-tuples of integers we say that $u \geq v$ if no component of u is less than the corresponding component of v.

> **Q**: Let X, Y be finite sets of m-tuples of integers. Is there an m-tuple z of integers such that the number of m-tuples x^i satisfying $x^i \geq z$ is at least as large as the number of m-tuples y^j satisfying $y^j \geq z$?

Reference: [PLA1].

12.9.7 Problems from Algebra and Number Theory

Quadratic Congruences Problem:

> **Q**: Let $a, b, c \in \mathbb{N}$. Is there a positive integer $x < c$ such that $x^2 \equiv a \bmod b$?

Reference: [MAA].

Comparative Divisibility Problem:

> **Q**: Let $\{a_j\}$ and $\{b_j\}$ be finite sequences of positive integers. Is there a positive integer c such that the number of indices i for which c divides a_i is more than the number of j for which c divides b_j?

Reference: [PLA1].

Exponential Expression Divisibility Problem:

> **Q**: Let $\{a_1, \ldots, a_n\}$ and $\{b_1, \ldots, b_m\}$ be finite sequences of positive integers. Let $q \in \mathbb{Z}$. Does $\prod_{i=1}^{n}(q^{a_i} - 1)$ divide $\prod_{j=1}^{m}(q^{b_j} - 1)$?

Reference: [PLA1].

Nontrivial Greatest Common Divisor Problem:

> **Q**: Let $p_1, \ldots, p_m$ be a collection of polynomials of one variable. Does the greatest common divisor of the p_j have degree greater than 0?

Reference: [PLA2].

Quadratic Diophantine Equations Problem:

> **Q**: Let $a, b, c \in \mathbb{N}$. Are there positive integers x, y such that $ax^2 + by^2 + c = 0$?

Reference: [MAA].

Root of Modulus One Problem:

> **Q**: Let p be a polynomial. Does p have a root on the complex unit circle?

Reference: [PLA2]. This problem is not even known to be in **NP**. We refer to such a problem as "**NP**-hard."

Cosine Product Integration:

> **Q**: Let $a_1, \ldots, a_n$ be integers. Is it the case that
>
> $$\int_0^{2\pi} \left(\prod_{i=1}^{n} \cos(a_i \theta) \right) d\theta = 0?$$

Reference: [PLA1].

12.9.8 Game and Puzzle Problems

Generalized Kayles Problem: Let $G = (V, E)$ be a graph. A game is played as follows. Players alternately choose a vertex in the graph, removing that vertex and all adjacent vertices. Player 1 wins precisely when Player 2 is the first player left with no vertices to choose.

> **Q**: Does Player 1 have a forced win in this game?

Reference: [SCHA2].

Sequential Truth Assignment: Given a finite collection of statements in the propositional calculus, players alternate assigning truth values to the variables. Player 1 wins precisely when the resulting truth assignments satisfy all the statements.

> **Q**: Does Player 1 have a forced win in this game?

Reference: [STM].

Alternating Hitting Set Problem: Let A be a given set and $\mathcal{C}$ a collection of subsets of A. A game is played as follows. Players alternate choosing a new element of A until, for each $c \in \mathcal{C}$, some element of c has been chosen. The player whose choice causes this to happen loses.

> **Q**: Does Player 1 have a forced win in this game?

Reference: [SCHA2].

$N \times N$ **Checker Problem**: A game of checkers is played on an $N \times N$ board. All of the pieces are assumed to be "kings." An initial position for the pieces is given.

Q: Does Black have a forced win from the given initial position?

Reference: [FGJSY].

12.9.9 Problems of Logic

The Satisfiability Problem:

Q: Let $\mathcal{C}$ be a collection of statements in the propositional calculus. Is there a truth assignment for the atomic statements that gives a satisfying truth assignment for the elements of $\mathcal{C}$?

Reference: [COO].

The Nontautology Problem:

Q: Let E be a statement in the propositional calculus. Is E *not* a tautology?

Reference: [COO].

Generalized Satisfiability Problem:

Q: Let $k_1, \ldots, k_m \in \mathbb{N}$. Let $S = \{R_1, \ldots, R_m\}$ be a sequence of subsets of $\{T, F\}^{k_i}$. Let U be a set of variables, and, for $i = 1, \ldots, m$, let C_i be a collection of k_i-tuples of variables from U. Is there a truth assignment $t : U \to \{T, F\}$ such that, for all i, $1 \leq i \leq m$ and for all k_i-tuples $(u_1, u_2, \ldots, u_{k_i}) \in C_i$, we have $(t(u_1), t(u_2), \ldots, t(u_{k_i})) \in R_i$?

Reference: [SCHA2].

12.9.10 Miscellaneous Problems

The Traveling Salesman Problem:

Q: There are k cities and a road between any two of them. The lengths of the roads are known. Find the shortest path that will enable the salesman to visit all of the cities and return to his starting point. [In the form of a decision problem: Given a positive number B, is there a path that is shorter than B?]

Reference: [COO].

Simply Deviated Disjunction Problem: Consider a collection M of k-tuples $(m_1^i, m_2^i, \ldots, m_k^i)$ with $i = 1, \ldots, n$. Here each m_j^i equals either 0 or 1 or x. We define Φ^i to be the disjunction over $j \in I$ of the statements $m_j^i = f(j)$ and Ψ^i to be the disjunction over $j \in I$ of the statements $m_j^i = f(j)$. We say that Φ and Ψ are *simply deviated in* M if [the number of $m^i \in M$ such that Φ^i and Ψ^i are both true] times [the number of $m^i \in M$ such that Φ^u and Ψ^i are both false] is larger than [the number of $m^i \in M$ such that Φ^i is true and Ψ^i is false] times [the number of $m^i \in M$ such that Φ^i is false and Ψ^i is true].

> **Q**: Let M be a collection of k-tuples $(m_1^i, m_2^i, \ldots, m_k^i)$, $i = 1, \ldots, n$; here each m_j^i equals either 0, 1, or x. Is there a partition of $\{1, 2, \ldots, k\}$ into sets $I \cup J$ and a function $f : \{1, 2, \ldots, k\} \to \{0, 1\}$ such that, if Φ^i is the disjunction over $j \in I$ of the statements $m_j^i = f(j)$ and Ψ^i is the disjunction over $j \in I$ of the statements $m_j^i = f(j)$, then $\{\Phi^i\}$ and $\{\Psi^i\}$ are simply deviated in M?

References: [HAV] and [PUS].

Matrix Cover Problem:

> **Q**: Let $F = (f_{ij})$ be an $n \times n$ matrix with nonnegative integer entries. Let $k \in \mathbb{Z}$. Is there a function $f : \{1, 2, \ldots, n\} \to \{-1, +1\}$ such that
>
> $$\sum_{1 \le i,j \le n} a_{ij} \cdot f(i) \cdot f(j) \ge k?$$

Reference: [GAJ, p. 282].

12.10 *More on* **P**/**NP**

We conclude this chapter by saying a few more words about the structure of **NP**. For details, we refer the reader to [GAJ, Chapter 7].

12.10.1 **NPC** *and* **NPI**

It is known that $\mathbf{P} \subseteq \mathbf{NP}$, and it is not known whether $\mathbf{P} = \mathbf{NP}$. We take as a working hypothesis the conjecture that most people believe: *we will assume in the rest of this discussion that* $\mathbf{P} \neq \mathbf{NP}$. Under that hypothesis, the following assertions are known to hold. Let **NPC** denote the **NP**-complete problems. Then $\mathbf{NPC} \subseteq [\mathbf{NP} \setminus \mathbf{P}]$. Moreover, it is known that $\mathbf{NPC} \neq [\mathbf{NP} \setminus \mathbf{P}]$. We let

$$\mathbf{NPI} \equiv [\mathbf{NP} \setminus \mathbf{P}] \setminus \mathbf{NPC}.$$

Then it is known that **NPI** is nonempty. The "I" in **NPI** denotes "intermediate," referring to "intermediate in difficulty."

12.10.2 Problems in NPI

On the one hand, it is not easy to name explicitly a problem that lies in **NPI**. On the other hand, **NPI** is known to contain infinitely many distinct equivalence classes of problems. Three problems that are believed to lie in **NPI**, although the final word has not been said about the matter, are:

The Graph Isomorphism Problem Let $G = (V, E)$ and $G' = (V', E')$ be graphs. Are G and G' combinatorially isomorphic?

The Composite Number Problem Let k be a positive integer. Does k factor as $k = m \cdot n$, with m and n integers greater than 1?

The Linear Programming Problem Let $V^i = (v_1^i, v_2^i, \ldots, v_n^i)$ be n-vectors for $i = 1, \ldots, m$. Let $D = (d_1, \ldots, d_m)$, $C = (c_1, \ldots, c_n)$, and let B be an integer. Is there a vector $X = (x_1, x_2, \ldots, x_n)$ such that $V^i \cdot X \leq d_i$ for $1 \leq i \leq m$ and also $C \cdot X \geq B$?

It is still an open problem to determine whether any of these problems is in **NPI**.

12.10.3 NP*-Hard Problems*

There are various refinements of the notion of **NP**-complete that have been studied in lieu of attacking **NP**-completeness head on. Any decision problem Π, whether a member of **NP** or not, to which we can transform an **NP**-complete problem will have the property that it cannot be solved in polynomial time unless $\mathbf{P} = \mathbf{NP}$. We call such a problem "**NP**-hard."

The notion of **NP**-hard problem has been formulated to include both decision problems and *search problems* that (whether they lie in **NP** or not) are equivalent to **NP**-complete problems. We refer the reader to [GAJ, Chapter 5] for further details about **NP**-hard problems and related ideas.

12.11 Descriptive Complexity Theory

Descriptive complexity theory is an aspect of finite model theory (Subsection 3.2.3). In its early days, finite model theory concentrated on finitistic versions of results from model theory taken as a whole. Since traditional model theory is about models of *any* cardinality, that approach is not very fruitful. Furthermore, the completeness and compactness theorems do not apply to finite structures. Fagin's theorem (Subsection 12.6.5) actually put finite model theory on the map. In the 1980's, Immerman proved that first-order logic extended with a least fixed point operation can express polynomial (and only polynomial) time properties of ordered, finite, first-order structures. These results laid the foundations of descriptive complexity theory. Descriptive set theory endeavors to find languages that describe, or characterize, complexity classes. Thus Immerman's theorem finds a language that characterizes polynomial complexity, at least in a certain context. Computer scientists endeavor to find languages that will carry out the goals of descriptive set theory. The article [MCA] gives more detail on the descriptive complexity theory program. See also [STE].

Chapter 13

Boolean Algebra

The application of whips, racks, gibbets, galleys, dungeons, fire and faggot in a dispute be looked upon as popish refinements upon the old heathen logic.

—Joseph Addison

Mathematics, rightly viewed, possesses not only truth, but supreme beauty—a beauty cold and austere, like that of sculpture, without appeal to any part of our weaker nature, without the gorgeous trappings of painting or music, yet sublimely pure, and capable of a stern perfection such as only the greatest art can show.

—Bertrand Russell

What is the task of all higher education? To make man into a machine. What are the means employed? He is taught how to suffer being bored.

—F.W. Nietzsche

There are no whole truths; all truths are half-truths. It is trying to treat them as whole truths that plays the devil.

—Alfred North Whitehead

One element of the soul is irrational and one has a rational principle.

—Aristotle

The general method which guides our handling of logical symbols is due to Peano. His great merit consists not so much in his definite logical discoveries nor in the details of his notations (excellent as both are), as in the fact that he first showed how symbolic logic was to be freed from its undue obsessions with the forms of ordinary algebra, and thereby made it a suitable instrument for research ...

—Bertrand Russell and Alfred North Whitehead

13.1 Description of Boolean Algebra

13.1.1 A System of Encoding Information

Boolean algebra, named after George Boole (1815–1864), is a formal system for encoding certain relationships that occur in many different logical systems. As an example, consider the algebra of sets, equipped with the operations $\cap$, $\cup$, ${}^c(\ )$. These are "intersection," "union," and "complementation." Propositional logic has three analogous operations: $\wedge$, $\vee$, $\sim$. If we think of these as corresponding,

$$\cap \longleftrightarrow \wedge$$

$$\cup \longleftrightarrow \vee$$

$${}^c(\ ) \longleftrightarrow \sim,$$

then we find that the two logical systems have very similar formulas:

$$\begin{aligned}
&\text{(i)}\ \left[{}^c(S \cup T) = {}^cS \cap {}^cT\right] \longleftrightarrow \\
&\qquad\qquad \left[\sim (A \vee B) \iff \sim A \wedge \sim B\right], \\
&\text{(ii)}\ \left[{}^c(S \cap T) = {}^cS \cup {}^cT\right] \longleftrightarrow \\
&\qquad\qquad \left[\sim (A \wedge B) \iff \sim A \vee \sim B\right], \\
&\text{(iii)}\ \left[S \cup (T \cap U) = (S \cup T) \cap (S \cup U)\right] \longleftrightarrow \\
&\qquad\qquad \left[A \vee (B \wedge C) \iff (A \vee B) \wedge (A \vee C)\right], \\
&\text{(iv)}\ \left[S \cap (T \cup U) = (S \cap T) \cup (S \cap U)\right] \longleftrightarrow \\
&\qquad\qquad \left[A \wedge (B \vee C) \iff (A \wedge B) \vee (A \wedge C)\right].
\end{aligned}$$

Other logical systems, such as the theory of gates in computer logic, or the theory of digital circuits in the basic theory of electricity, satisfy analogous properties. Boolean algebra abstracts and unifies all of these ideas into a single algebraic system.

The *Stone Representation Theorem* [JOH] states that every Boolean algebra can be realized as the algebra of closed-and-open sets on a compact, zero-dimensional topological space. This unifies several different areas of mathematics and has proved to be a powerful point of view.

13.2 Axioms of Boolean Algebra

13.2.1 Boolean Algebra Primitives

Boolean algebra contains these primitive elements: A collection, or set, S of objects. At a minimum, S will contain the particular elements 0 and 1. Boolean algebra also contains three operations (two binary and one unary): $+$, $\times$, and $^{-}$. Boolean algebra uses the equal sign $=$ and parentheses (,) in the customary manner. In Boolean algebra, just as in fuzzy set theory, we think of the overbar as denoting set complementation (although it could have other specific meanings in particular contexts).

13.2.2 Axiomatic Theory of Boolean Algebra

The axioms for Boolean algebra, using elements $a, b, c \in S$, are these:

1. $a \times 1 = a$;
2. $a + 0 = a$;
3. $a \times b = b \times a$;
4. $a + b = b + a$;
5. $a \times (b + c) = (a \times b) + (a \times c)$;
6. $a + (b \times c) = (a + b) \times (a + c)$;
7. $a \times a = 1$;
8. $a \times \overline{a} = 0$;
9. $a + \overline{a} = 1$.

Some of these axioms (1, 2, 3, 4, 5) have a familiar form that we have seen in ordinary arithmetic. The other four are not familiar, and may be counter-intuitive. In fact axioms 6, 7, 8, 9 are *false* in ordinary arithmetic. So Peano's arithmetic is *not* a model for Boolean algebra. But the algebra of sets *is* a model for Boolean algebra. If we let S be the collection of all sets of real numbers, and if we interpret $+$ as union ($\cup$), $\times$ as intersection ($\cap$), $^{-}$ as set-theoretic complementation ($^{c}(\ )$), 0 as the empty set ($\emptyset$), and 1 as the universal set (in this model, the real numbers $\mathbb{R}$), then in fact all nine axioms now hold. As an example,

$$S \cap {}^{c}S = \emptyset$$

is the correct interpretation of axiom 8, and it is now true. Also

$$S \cup (T \cap U) = (S \cup T) \cap (S \cup U)$$

is the correct interpretation of axiom 6, and it is now true. Similar statements may be made about the interpretation of the axioms of Boolean algebra in the propositional calculus.

13.2.3 Boolean Algebra Interpretations

Obversely, we can also give the Boolean algebra interpretations of the four sets of equivalent statements that we gave in Section 13.1.1. These are

(*i*) $$\overline{(a + b)} = \overline{a} \times \overline{b},$$

(*ii*) $$\overline{(a \times b)} = \overline{a} + \overline{b},$$

(*iii*) $$a + (b \times c) = (a \times b) + (a \times c),$$

(*iv*) $$a \times (b + c) = (a \times b) + (a \times c).$$

13.3 Theorems in Boolean Algebra

13.3.1 Properties of Boolean Algebra

One of the remarkable features of Boolean algebra is that it has a very small set of axioms, yet many additional desirable properties are readily derived. Some of these properties are:

(10) $a \times a = a$;

(11) $a + a = a$;

(12) $a \times 0 = 0$;

(13) $a + 1 = 1$;

(14) $0 \neq 1$;

(15) 0 is unique and 1 is unique;

(16) $\overline{\overline{a}} = a$;

(17) $a + (b + c) = (a + b) + c$;

(18) $a \times (b \times c) = (a \times b) \times c$;

(19) $\overline{(a \times b)} = \overline{a} + \overline{b}$;

(20) $\overline{(a + b)} = \overline{a} \times \overline{b}$;

(21) $\overline{0} = 1$;

(22) $\overline{1} = 0$;

(23) $a \times (a + b) = a$;

(24) $a + (a \times b) = a$;

(25) $\overline{a} \times (a \times b) = 0$;

(26) $\overline{a} + (a + b) = 1$;

(27) $a \times (\overline{a} \times \overline{b}) = 0$;

(28) $a + (\overline{a} + \overline{b}) = 1$;

(29) If, $a \times c = b \times c$ and $a + c = b + c$, then $a = b$.

13.3.2 A Sample Proof

As an illustration of the ideas, we now provide a proof of formula (11). The proof consists of a sequence of statements, each justified by one of the nine axioms of Boolean algebra or by the rules of logic. At each step, we cite the relevant axiom or rule.

1. [**Axiom (2)**] $a + 0 = a$;
2. [**Axiom (4)**] $a = a + 0$;
3. [**Axiom (7)**] $a \times \overline{a} = 0$;
4. [**Axiom (3)**] $0 = a \times \overline{a}$;
5. [**Steps** 2 **and** 4] $a = a + (a \times \overline{a})$;
6. [**Axiom (6)**] $a = (a + a) \times (a + \overline{a})$;
7. [**Axiom (8)**] $a = (a + a) \times 1$;
8. [**Axiom (1) + Step** 4 **+ Step** 6] $a = a + a$;
9. [**Symmetry of equality**] $a + a = a$.

Some of the elementary theorems of Boolean algebra that we have cited require rather elaborate justifications. For instance, it is rather difficult to prove the associativity of addition (Theorem 17). We refer the reader to [NIS] for the details.

13.4 Illustration of the Use of Boolean Logic

We now present an example adapted from one that is presented in [NIS, p. 41]. Imagine an alarm system for the monitoring of hospital patients. There are four factors (inputs/outputs) that contribute to the triggering of the alarm:

Inputs/Outputs	**Meaning**
a	Patient's temperature is in the range 36–40° C.
b	Patient's systolic blood pressure is outside the range 80–160 mm.
c	Patient's pulse rate is outside the range 60–120 beats per minute.
o	Raising the alarm is necessary.

Good sense dictates that we would want the alarm to sound if any of the following situations obtains:

- The patient's temperature is outside the acceptable range, the blood pressure is outside the acceptable range, and the pulse rate is outside the acceptable range.
- The patient's temperature is in the acceptable range, but the pulse rate is outside the acceptable range.
- The patient's temperature is in the acceptable range, but the systolic pressure is outside the acceptable range.
- The patient's temperature is in the acceptable range, but both the systolic pressure and the pulse rate are outside the acceptable range.

13.4.1 Boolean Algebra Analysis

Using Boolean algebra, these four conditions can be encoded as

- $\overline{a} \times b \times c$;
- $a \times \overline{b} \times c$;
- $a \times b \times \overline{c}$;
- $a \times b \times c$.

Notice that, since we know that multiplication is associative, we have omitted using parentheses to group these binary operations. No ambiguity results.

The aggregate of all the situations in which we want the alarm to sound can be represented by the equation

$$o = [\overline{a} \times b \times c] + [a \times \overline{b} \times c] + [a \times b \times \overline{c}] + [a \times b \times c]. \qquad (\dagger)$$

Now we can use the laws of Boolean algebra to simplify the right-hand side:

$$\begin{aligned}
&[\overline{a} \times b \times c] + [a \times \overline{b} \times c] + [a \times b \times \overline{c}] + [a \times b \times c] \\
&= [\overline{a} \times b \times c] + [a \times \overline{b} \times c] + [a \times b \times \overline{c}] \\
&\quad + \Big([a \times b \times c] + [a \times b \times c] + [a \times b \times c]\Big) \\
&= \Big([\overline{a} \times b \times c] + [a \times b \times c]\Big) \\
&\quad + \Big([a \times \overline{b} \times c] + [a \times b \times c]\Big)
\end{aligned}$$

$$
\begin{aligned}
&+\Big([a \times b \times \overline{c}] + [a \times b \times c]\Big) \\
&= \Big([\overline{a} + a] \times [b \times c]\Big) \\
&\quad +\Big([\overline{b} + b] \times [a \times c]\Big) \\
&\quad +\Big([\overline{c} + c] \times [a \times b]\Big) \\
&= \Big(1 \times [b \times c]\Big) + \Big(1 \times [a \times c]\Big) + \Big(1 \times [a \times b]\Big) \\
&= [b \times c] + [a \times c] + [a \times b].
\end{aligned}
$$

In summary, we have reduced our protocol for the sounding of the medical alarm to:

$$o = [b \times c] + [a \times c] + [a \times b]. \qquad (\ddagger)$$

The circuit that we put in place can be designed after equation (‡) rather than after the much more complicated equation (†).

Chapter 14

The Word Problem

Logical consequences are the scarecrows of fools and the beacons of wise men.

—Thomas Henry Huxley

It is undesirable to believe a proposition when there is no ground whatever for supposing it to be true.

—Bertrand Russell

Logic is a large drawer, containing some useful instruments, and many more that are superfluous.

—C.C. Colton

How can one talk about 'understanding' a proposition and 'not understanding' a proposition? Surely it's not a proposition until it's understood?

—Ludwig Wittgenstein

The philosophy of reasoning, to be complete, ought to comprise the theory of bad as well as of good reasoning.

—J.S. Mill

Mathematical induction often arises as the finishing step ... of an inductive research.

—George Polya

Multiplication is vexation,
Division is as bad; The rule of three doth puzzle me,
And practice drives me mad.

—A Nursery Rhyme

There is something in the vanity of logic which addles a man's brains.

—E.A. Poe

Philosophy is the science which considers truth.

—Aristotle

We have avoided both controversy and general philosophy, and made our statements dogmatic in form ...

—Bertrand Russell and Alfred North Whitehead

14.1 Introductory Remarks

The *word problem* is a decision problem in group theory. It turns out to be a problem that is formally undecidable. We will explain here the necessary background in group theory and what the word problem is, and we will provide some discussion of the undecidability issue.

14.2 What Is a Group?

A *group* is a set G that is equipped with a binary operation, which we think of as multiplication and denote by $\cdot$. If $g, h \in G$, then we assume that $g \cdot h \in G$ (this is the *closure property* of the group operation). The axioms for a group are:

(1) The group operation is associative: If $g, h, k \in G$, then $g \cdot (h \cdot k) = (g \cdot h) \cdot k$.

(2) There is an element $e \in G$ such that $e \cdot g = g \cdot e$ for every $g \in G$.

(3) For every element $g \in G$ there is an element $h \in$G such that $g \cdot h = h \cdot g = e$. We commonly denote this element h by g^{-1}.

14.2.1 First Consequences

It is straightforward to check, directly from the definitions and the axioms, that the identity element e is unique. Also the inverse of each element of the group (Axiom (3)) is unique. We do *not* assume, nor is it true in general, that the group is commutative (i.e., *abelian*). That is, we do not assume that $g \cdot h = h \cdot g$. In fact, the word problem to be discussed in this chapter is trivial, and certainly decidable, for abelian groups.

14.2.2 Subgroups and Generators

A subset $H \subseteq G$ is called a *subgroup* if (i) $e \in H$, (ii) whenever $g, h \in H$, then $g \cdot h \in H$, and (iii) whenever $g \in H$, then $g^{-1} \in H$. If $B \subseteq G$ is any set (not necessarily a subgroup), then we let $\{B\}$ denote the *subgroup* generated by B. Thus $\{B\}$ consists of all (finite) expressions of the form

$$b_1^{\pm 1} \cdot b_2^{\pm 1} \cdots b_k^{\pm 1}$$

formed with elements $b_j \in B$.

14.2.3 Homomorphisms

If G and H are any groups, then a *homomorphism* $\phi : G \to H$ from G to H is a function with the property that $\phi(g \cdot h) = \phi(g) \cdot \phi(h)$.

If the homomorphism is both one-to-one and onto, then we call it an *isomorphism.*

We call $\{e\}$ the *zero subgroup* of G. An element $g \in G$ is called *nonzero* if it is different from e. If $\phi : G \to H$ is a mapping or a homomorphism and if $B \subset G$ is a subset, then we let $\phi\big|_B$ denote the *restriction* of ϕ to B; this means that we just consider B to be the domain of ϕ.

14.3 What Is a Free Group?

14.3.1 The Definition

The *free group* $[A]$ on a set A is a group including A that has the following property: Every element of $[A]$ can be uniquely written in the form

$$a_1^{\pm 1} \cdot a_2^{\pm 1} \cdots a_k^{\pm 1}, \qquad (*)$$

where the a_i are in A and and no a_j appears adjacent to a_j^{-1}. (Note that we allow $k = 0$; this gives the expression for e in the free group.) The expressions $(*)$ are called *words* on A. Notice that words, and the free group, may be formed from A even if A is an arbitrary set; A need *not* be a subset of some given group. The words are the elements of the free group $[A]$.

14.3.2 Words

We multiply two words by juxtaposing them (in some order) and then crossing out any expressions $a \cdot a^{-1}$ or $a^{-1} \cdot a$ that may result from that juxtaposition. After all of these crossings out are performed (a process called *reduction*), we obtain a new word that is the product of the original two words. If A is a subset of a given group G, then $\{A\} = [A]$.

14.4 The Word Problem

Our treatment of the word problem draws significantly on the elegant exposition in [SCH].

14.4.1 Extensions of Homomorphisms

If A is a set and if ϕ is a mapping (not necessarily a homomorphism) of A into a group G, then there is a unique extension $\widetilde{\phi}$ from $[A]$ into G that *is* a homomorphism. Observe that $\widetilde{\phi}$ is surjective if and only if $\phi(A)$ generates G. In particular, if A is a generating set for a group G, then there is a unique homomorphism from $[A]$ to G that extends the identity from A to G. *Note that the extended homomorphism may not be the identity.* This extended homomorphism will be surjective. By the first fundamental isomorphism of algebra (if $\phi : L \to M$ is surjective then $M \cong L/\ker\phi$), the group G is therefore isomorphic to a factor group of $[A]$. As a corollary, we note that every finitely generated group is a quotient, or factor group, of a free group on a finite set.

14.4.2 An Illustrative Example

Example 14.1

Let G be the group of rotations of $\mathbb{R}^3$ that preserve the unit cube in $\mathbb{R}^3$. Let ϕ be counterclockwise rotation by angle $\pi/2$ in the x-y plane, let ψ be counterclockwise rotation by angle $\pi/2$ in the y-z plane, and let ρ be counterclockwise rotation by angle $\pi/2$ in the x-z plane. Write $A = \{\phi, \psi, \rho\}$. Then the map

$$A \to G$$

given by inclusion extends to a homomorphism of $[A]$ into G. But the extended map is clearly not the identity. Indeed, the elements $\phi^2\rho^2\psi^{-2}$, $\phi^2\psi^2\rho^{-2}$, and $\rho^2\psi^2\phi^{-2}$ are all in the kernel of the mapping.

Thus we see that G is the quotient of the free group on three letters by the kernel of this mapping.

□

A subset A of a group G is *free* if the homomorphism from $[A]$ to G that is the identity on A is injective. In this case, we may identify the subgroup $\{A\}$ of G with $[A]$. We shall also say in this circumstance that the subset A *has no relations*. We will say more about relations below.

14.5 Relations and Generators

A relation on a set $A \subset G$ is an expression of the form $X = Y$, where X and Y are words on A. This relation is said to *hold* in a factor group $[A]/K$ if X and Y lie in the same coset of K, equivalently, if $XY^{-1} \in K$.

14.5.1 Consequences

Suppose that R is a collection of relations on A. A relation on A is a *consequence* of R if it holds in every factor group of $[A]$ in which all the relations of R hold. The set of consequences of R is denoted by $\mathcal{C}(R)$. Let K_R be the normal subgroup[1] generated by all expressions of the form $X \cdot Y^{-1}$ for any X, Y with $X = Y$ in R. The factor group $[A]/K_R$ is precisely the one in which the relations that hold are the consequences

[1]A subgroup H in G is said to be normal if $g^{-1}hg \in H$ whenever $g \in G$ and $h \in H$.

of R. We call $[A]/K_R$ the group with *generators* A and *defining relations* R. We designate that group by $[A; R]$.

14.5.2 Generators and Relations

We will amplify the notation $[A; R]$ to allow for several generators or sets of generators before the semicolon and several relations or sets of relations after the semicolon. Thus, for example, $[A, x; R, X = Y]$ has the set of generators $A \cup \{x\}$ and the set of defining relations $R \cup \{X = Y\}$.

We build a certain nonredundancy into our ideas. It is understood that no generator is repeated. In this last example, it must therefore be that $x \notin A$. If A appears before the semicolon and a appears after the semicolon, then it is understood that a varies over A. For instance, in $[A, x; ax = xa]$, the defining relations are all the expressions $ax = xa$ for $a \in A$.

14.6 Amalgams

Let A and B be subgroups of a group G. Then $[A]$ is a subgroup of the group $[A, B]$, and if R is a set of relations on A and S is a set of relations on B, then K_R is a subgroup of $K_{R\cup S}$. Thus there is a natural homomorphism from $[A; R]$ to $[A, B; R, S]$ that maps the coset in $[A; R]$ of a word on A into the coset of that same word in $[A, B; R, S]$. If that homomorphism is bijective, then we identify those two groups. This situation is of particular interest in the case when S consists of one relation $b = X$ with X a word on A for each $b \in B$.

Based on our earlier discussion, we know that a group G is naturally isomorphic to a factor group of $[G]$, and hence to the group $[G; R_G]$, where R_G is the set of all relations on G that hold in the factor group. When G appears before the semicolon, this is understood to signify that the relations in R_G are among the defining relations, even if they do not appear explicitly after the semicolon.

14.6.1 Free Product with Amalgamation

Let G and G' be groups. Let ϕ be an isomorphism of a subgroup H of G with a subgroup H' of G'. The *free product* of the groups G and G' with the *amalgamation* ϕ is the group

$$G *_\phi G' \equiv [G, G'; h = \phi(h)].$$

The natural mappings of G and G' into $G *_\phi G'$ are injective; we therefore identify G and G' with their images under these natural mappings. Then H and H' are also identified via the isomorphism ϕ. We have that

$G *_\phi G' = \{G, G'\}$ and $G \cap G' = H = H'$. This last group is called the *amalgam.*

Let T consist of one element from each right coset of H in G other than H itself, and let T' be formed similarly from G' and H'. A word on $G \cup G'$ is said to be in *normal form* if it can be written as $ht_1t_2 \cdots t_n$, where $h \in H$ and $t_1, \ldots, t_n \in T \cup T'$ and $t_i \in T$ if and only if $t_{i+1} \in T'$, $1 \leq i < n$. Schreier's theorem states that every coset in $G *_\phi G'$ contains exactly one word in normal form.

Let K and K' be subgroups of G and G', respectively, so that

$$\phi(H \cap K) = \phi(H) \cap K.$$

[Thus, in $G *_\phi G'$;, K and K' have the same intersection with the amalgam.] The restriction $\psi = \phi\big|_{H \cap K}$ is then an isomorphism of $H \cap K$ and $\phi(H) \cap K$. We thus may form $K *_\psi K'$; and there is a natural mapping χ from $K *_\psi K'$ to $G *_\phi G'$ whose image is $\{K, K'\}$. It turns out that χ is injective, so we may identify $K * \psi K'$ with $\{K, K'\} \subset G *_\phi G'$. Then it may be shown that

$$G \cap \{K, K'\} = K.$$

14.6.2 The Free Product

In the special case where ϕ is an isomorphism of the zero subgroups, we write $G * G'$ for $G *_\phi G'$ and we call $G * G'$ the *free product* of G and G'. Thus $G * G' = [G, G']$. The normal forms in this case become the products of nonzero elements that are alternately in G and G' (assuming that we omit the initial e). The K and K' described above can, in this instance, be any subgroups of G and G'.

Now let G and G' be any subgroups of a group L. Let $H = G \cap G'$. Let ϕ be the identity mapping from H to H. Then, by the theory developed above, there is a unique homomorphism from $G *_\phi G'$ to L that is the identity on G and G'. The image of this homomorphism is $\{G, G'\}$. In case this homomorphism is injective, we identify $G *_\phi G'$ with $\{G, G'\}$ and we say that $\{G, G'\}$ is the *free product* of G and G' with the amalgam H. (We omit explicit mention of H in the special case where H is the zero subgroup.)

14.6.3 Finitely Presented Groups

We say that a group is *finitely presented* if it is isomorphic to a group $[A; R]$ and both A and R are finite sets. Clearly, the direct product of finitely presented sets is finitely presented. The free product of two finitely presented groups with a finitely presented amalgam is also finitely presented.

14.7 Description of the Word Problem

Let R be a set of relations on a finite set A. The *word problem* for R is the decision problem for $\mathcal{C}(R)$. Since $\mathcal{C}(R)$ is the set of relations holding in $[A; R]$, we also call this problem the *word problem for* $[A; R]$.

14.7.1 The Word Problem and Recursion Theory

It is useful to translate the word problem into the language of recursion theory. To this end, we identify the symbols a and a^{-1} (for $a \in A$) and also the symbol $=$ with natural numbers. A word or relation on A then is translated to a finite sequence of natural numbers. Thus the word or relation will be given a sequence number (much as in the theory of Gödel numbers). A set $\mathbf{P}$ of words or relations is then said to be *recursive* (or *recursively enumerable*) if the set of sequence numbers of elements of $\mathbf{P}$ is recursive (or recursively enumerable). Clearly, this notion is independent of the protocol used to identify words and relations with numbers. The word problem for R is thus solvable if and only if $\mathcal{C}(R)$ is recursive.

A relation of the form $X = Y$ is in $\mathcal{C}(R)$ if and only if $X \cdot Y^{-1}$ lies in K_R. And X is in K_R if and only if $X = e$ is in $\mathcal{C}(R)$. Therefore K_R is recursive (or recursively enumerable) if and only if $\mathcal{C}(R)$ is recursive (or recursively enumerable). Furthermore, if R is recursively enumerable, then $\mathcal{C}(R)$ is recursively enumerable.

14.7.2 Recursively Presented Groups

A group is *recursively presented* if it is isomorphic to a group $[A; R]$, where A is finite and R is recursively enumerable. For example, every finitely presented group is recursively presented.

Assume that A and B are finite. Suppose that $[A; R]$ is embedded in $[B; S]$. If the word problem for $[B; S]$ is solvable, then it can be shown that the word problem for $[A; R]$ is solvable. Also, if $\mathcal{C}(S)$ is recursively enumerable, then $\mathcal{C}(R)$ is recursively enumerable. Since $[A; R] = [A; \mathcal{C}(R)]$, it follows that a subgroup of a recursively presented group is recursively presented.

14.7.3 Solvability of the Word Problem

Now we come to the main point of this discussion. If G is a finitely presented group, then G is isomorphic to a group $[A; R]$ with A finite. We say that the word problem for G is *solvable* or *unsolvable* according to whether the word problem for $[A; R]$ is solvable or unsolvable. This result is independent of the choice of the representation $[A; R]$.

One of the first main results of this subject, and the reason for the inclusion of this material in the present book, is the following.

14.7.4 Novikov's Theorem

Theorem 14.1 (Novikov)
There is a finitely presented group that has an unsolvable word problem.

Thus we have been able to say, in a very precise sense, that the word problem is undecidable. There are many problems in modern mathematics, some of them having to do with the construction of invariants for geometric structures, others having to do with cryptography or coding theory, and still others remaining to be unearthed, that are equivalent to the word problem. The word problem is a powerful weapon for seeing that certain types of problems are formally undecidable.

List of Notation from Logic

Symbol	**Meaning**
wff	well-formed formula
$\wedge$	and
$\vee$	or
$\sim$	not
$\Rightarrow$	implies
$\Longleftrightarrow$	if and only if
$\mid$	nand
$\downarrow$	nor
$\oplus$	exclusive "or"
$\Box$	it is necessary that
$\diamond$	it is possible that
$\forall$	for all
$\exists$	there exists
$\text{Consis}\, T$	T is consistent
$\models \varphi$	φ is a tautology
$S \models \varphi$	S tautologically implies φ
$S \vdash \varphi$	S proves φ
$=$	equal sign
$S = \{x \in \mathbb{R} : x \geq 3\}$	set-builder notation

Symbol	Meaning
$a_j b^j$	Einstein summation notation
$c_\alpha \mapsto \overline{c_\alpha}$	interpretation in a model
$R_\beta \mapsto \overline{R_\beta}$	interpretation in a model
$\exists!$	there exists a unique
Z_1	a formulation of arithmetic
Z_2	a formulation of arithmetic
p.r.	primitive recursive
AC	Axiom of Choice
ZF	Zermelo–Fraenkel set theory
ZFC	Zermelo–Fraenkel set theory with the Axiom of Choice
CH	the continuum hypothesis
$\mathcal{K}(A)$	the cardinality of A
$\mathcal{P}(A)$	the power set of A
$A \setminus B$	set-theoretic difference of A and B
$\mathbb{R}$	the real numbers
$\mathbb{Q}$	the rational numbers
$\mathbb{N}$	the natural numbers
$\mathbb{Z}$	the integers
$S \times T$	product of sets S and T
$\emptyset$	the empty set
(s, t)	ordered pair
$\in$	element of
$\notin$	not an element of
$A \subset B$	A is a subset of B
$A \subseteq B$	A is a subset of B
$\underset{\neq}{\subset}$	is a subset of but not equal to

Symbol	Meaning
$\neq$	not equal to
$\nsubseteq$	not a subset of
$\not\subset$	not a subset of
$A \cup B$	A union B
$A \cap B$	A intersect B
${}^{c}A$	complement of A
$\mathcal{R}$	a relation
$x\mathcal{R}y$	x is related to y
$\bigcup_{\alpha \in A} S_\alpha$	union over the index set A of S_α, $\alpha \in A$
$\bigcap_{\alpha \in A} S_\alpha$	intersection over the index set A of S_α, $\alpha \in A$
$\prod_{\alpha \in A} S_\alpha$	product over the index set A of S_α, $\alpha \in A$
$S \triangle T$	symmetric difference of S and T
E	the set of even integers
O	the set of odd integers
$x \sim y$	x is related to y
$[x]$	the equivalence class containing x
$\mathbb{C}$	the complex numbers
$\mathbb{H}$	the quaternions
$\mathbb{O}$	the Cayley numbers
f	a function
$f(x)$	the value of the function f at x
$g : X \to Y$	a function from X to Y
$g : x \mapsto y$	g maps x to y
$\|S\|$	the cardinality of S
$\mathbb{Q}^p$	the positive rational numbers

Symbol	Meaning
$\mathbb{R}^+$	the nonnegative real numbers
$\mathbb{Z}^+$	the nonnegative integers
$\mathbb{Q}^+$	the nonnegative rational numbers
ω	the cardinality of $\mathbb{N}$
$\aleph_0$	aleph-nought; the cardinality of $\mathbb{N}$
c	the cardinality of the continuum
$\aleph_1$	aleph-one; the least cardinality that exceeds $\aleph_0$
$\aleph_n$	the nth cardinality
MA	Martin's Axiom
$(P, \leq)$	a partially ordered set
2^ω	the cardinality of $\mathcal{P}(\mathbb{N})$
$\mathbf{m}$	a cardinal number
ccc	countable chain condition
sup	supremum
x'	the successor of x
$\text{Lim}(\alpha)$	α is a limit ordinal
$<_o$	partial ordering in the ordinal numbers
$\leq_o$	partial ordering in the ordinal numbers
χ_S	the characteristic function of S
(X, A)	a fuzzy set
f_y^b	the indicator function of a fuzzy point
F, G	the operator functions for fuzzy set theory
$\overline{A}$	complementation in fuzzy set theory
$T(a, b)$	a triangular norm in fuzzy set theory
$S(a, b)$	a triangular conorm in fuzzy set theory
λ (variable) $\bullet$ (function_body)	syntax for a Lambda-function

Symbol	**Meaning**
λ (signature) $\mid$ (constraint) $\bullet$ (function_body)	syntax for a Lambda-function
$(M\,N)$	application of M to N in the λ-calculus
$(\lambda x \bullet M)$	an abstraction in the λ-calculus
$E[x/M]$	notation for substitution in the λ-calculus
$\#$	counting operation in bag theory
count	counting operation in bag theory
$\uplus$	union operation in bag theory
$\bigcup$	difference operation in bag theory
E	element of in bag theory
$\sqsubseteq$	subset of in bag theory
i	the square root of -1
$\mathbf{i}, \mathbf{j}, \mathbf{k}$	basis elements for the quaternions
$\mathbf{i}_0, \mathbf{i}_1, \ldots, \mathbf{i}_6$	basis elements for the Cayley numbers
$\mathbb{R}^*$	the nonstandard real numbers
$P(j)$	an inductive statement
$\mathcal{U}$	a universe
$\text{Set}_{\mathcal{U}}$	the sets of a universe $\mathcal{U}$
$\text{Fn}_{\mathcal{U}}$	the functions of a universe $\mathcal{U}$
$\sigma, \tau \mapsto \sigma \times \tau$	a formal operation in determining type
$(\sigma \to \tau)$	a formal operation in determining type
$[\sigma]$	a formal operation in determining type
$\perp$	the false or empty clause
$\top$	the true clause

Symbol	Meaning
$L(A)$	the nonlogical symbols plus free variables in A
$\mathrm{dom}\, f$	the domain of f
$\mathrm{cod}\, f$	the codomain of f
id_a	the identity arrow on a
$f \circ g$	composition of f and g
$A \times_{\mathcal{O}} A$	product over $\mathcal{O}$
$\hom(b, c)$	the set of arrows from b to c
$\downarrow\downarrow$	the category with two objects and two arrows
$H_n(X)$	the nth singular homology group of X
τ	a natural transformation
class $\mathbf{P}$	problems of polynomial complexity
class $\mathbf{NP}$	problems with polynomial verification time for a guess
class $\mathbf{NPC}$	$\mathbf{NP}$-complete problems
class $\mathbf{NPI}$	$[\mathbf{NP} \setminus \mathbf{P}] \setminus \mathbf{NPC}$
G	a group
e	the identity element in a group
g^{-1}	the inverse of an element g of a group
$\{B\}$	subgroup generated by B
$[A]$	free group on a set A
$a_1^{\pm 1} \cdot a_2^{\pm 1} \cdots a_k^{\pm 1}$	element of a free group
$\mathcal{C}(R)$	the set of consequences of the relations in R
K_R	the normal subgroup generated by $X \cdot Y^{-1}$ for $X = Y$ in the given relations
$G *_{\phi} G'$	free product of the groups G, G' with the amalgamation ϕ

Glossary of Terms from Mathematical and Sentential Logic

abelian A group is abelian if it is *commutative.*

algebraic closure Let k be a field. The algebraic closure of k is the smallest field $\widetilde{k}$ containing k with the property that any polynomial with coefficients in $\widetilde{k}$ has all of its roots in $\widetilde{k}$.

algebraic theory An algebraic theory T is a category whose objects are the natural numbers $\mathbb{N}$ and that for each $n \in \mathbb{N}$ is equipped with an n-tuple of maps

$$\text{proj} : n \to 1, \quad i = 1, \ldots, n;$$

this makes n into the n-fold categorical product of 1, $n = 1^n$.

amalgam See also *free product with amalgamation.* Let H be a subgroup of the group G and H' a subgroup of the group G'. Let ϕ be an isomorphism of H to H'. The natural mappings of G and G' into $G *_\phi G'$ are injective; we therefore identify G and G' with their images under these natural mappings. Then H and H' are also identified via the isomorphism ϕ. We have that $G *_\phi G' = \{G, G'\}$ and $G \cap G' = H = H'$. This last group is called the amalgam.

amalgamation See also *free product with amalgamation.*

and The logical connective that conjoins two component statements, denoted by $\wedge$. If $\mathbf{A}$ or $\mathbf{B}$ are both true, then $\mathbf{A} \wedge \mathbf{B}$ is true. The statement $\mathbf{A} \vee \mathbf{B}$ is false if either $\mathbf{A}$ or $\mathbf{B}$ or both are false.

arrow In category theory, an arrow is a function or mapping of objects.

arrow function An operation that assigns to each arrow $f : c \to c'$ of C an arrow $Tf : Tc \to Tc'$ of B so that

$$T(\mathbf{1}_c) = \mathbf{1}_{Tc}$$
$$T(g \circ f) = Tg \circ Tf \text{ whenever } g \circ f \text{ is defined in } C.$$

Artin–Schreier theorem Any field in which -1 is not a sum of squares can be ordered.

associative A group operation $\cdot$ is said to be associative if $g \cdot (h \cdot k) = (g \cdot h) \cdot k$ for elements g, h, k in the group G. There are similar definitions of "associative" for set-theoretic union, set-theoretic intersection, and other contexts as well.

atomic statement An atomic statement is a sentence with a subject or a verb but no connectives; more specifically, an expression $Pt_1 \cdots t_n$ (or $P(t_1, \ldots, t_n)$), with $t_1, \ldots, t_n$ terms and $\mathbf{P}$ an n-place predicate symbol.

axiom A statement whose truth is taken for granted as part of a logical system. Usually an axiom is formulated in such a way that its validity is intuitive and self-evident. There are both logical axioms and nonlogical axioms. Also called a *postulate*.

Axiom for Cardinals An axiom of Zermelo–Fraenkel set theory.

Axiom for Existence An axiom for the predicate calculus.

Axiom for Prenex Normal Form An axiom for the predicate calculus.

Axiom of Choice An axiom of Zermelo–Fraenkel set theory. Perhaps the most bewildering of all the axioms of set theory, this axiom simply asserts that, from any collection of sets, one element may be selected from each.

Axiom of Extensionality An axiom of Zermelo–Fraenkel set theory.

Axiom of Inaccessible Cardinals An additional axiom for Zermelo–Fraenkel set theory that postulates the existence of certain large cardinals.

Axiom of Infinity An axiom of Zermelo–Fraenkel set theory.

Axiom of Regularity An axiom of Zermelo–Fraenkel set theory.

Axiom of Separation See the Axiom Schema of Replacement.

Axiom Schema of Replacement An axiom of Zermelo–Fraenkel set theory.

bag A set in which multiple occurrences of objects are allowed.

Banach–Tarski paradox Let B be the closed unit ball in $\mathbb{R}^3$. There is a decomposition

$$B = S \cup T$$

such that $S \cap T = \emptyset$, $S = S_1 \cup S_2 \cup \cdots \cup S_k$, $T = T_1 \cup T_2 \cup \ldots \cup T_k$, $U = U_1 \cup U_2 \cup \cdots \cup U_k$ and, for each j, S_j, T_j, and U_j are geometrically congruent.

binary relation See *k-ary relation, 2-ary relation.*

Boolean algebra Boolean algebra contains these primitive elements: A collection, or set, S of objects. At a minimum, S will contain the particular elements 0 and 1. Boolean algebra also contains three operations (two binary and one unary): $+$, $\times$, and $-$. Boolean algebra uses the equal sign ($=$) and parentheses ((,)) in the customary manner.

The axioms for Boolean algebra, using elements $a, b, c \in S$, are:

1. $a \times 1 = a$;
2. $a + 0 = a$;
3. $a \times b = b \times a$;
4. $a + b = b + a$;
5. $a \times (b + c) = (a \times b) + (a \times c)$;
6. $a + (b \times c) = (a + b) \times (a + c)$;
7. $a \times a = 1$;
8. $a \times \overline{a} = 0$;
9. $a + \overline{a} = 1$.

bound and free variables in the Lambda-calculus Let x be a variable in a λ-term M. We say that x is a *bound variable* if it occurs within a part of M having the form $\lambda x \bullet N$. Otherwise we say that the occurrence of x is free.

bound variable A variable of a function that satisfies one of the following stipulations:

(i) Variables in $\sim \mathbf{A}$, $\mathbf{A} \wedge \mathbf{B}$, $\mathbf{A} \vee \mathbf{B}$, $\mathbf{A} \Rightarrow \mathbf{B}$, and $\mathbf{A} \iff \mathbf{B}$ are bound according to whether they are bound in **A** or **B** separately.

(ii) The bound occurrences of a variable in a formula of the form $\exists x, \mathbf{A}$ or $\forall x, \mathbf{A}$ are the same as the bound occurrences of the variable itself, *except* that every occurrence of X is now considered bound.

Cantor diagonalization argument A method of proof that is used to show that certain sets are uncountable.

Cantor's theorem Let S be any set. Then the set $\mathcal{P}(S)$ of all subsets of S has larger cardinality than S.

cardinality See also *cardinal number.*

cardinal number An equivalence class of sets, all of which are set-theoretically isomorphic to each other.

category A category is a graph with two additional functions:

$$\begin{aligned} O &\overset{\text{id}}{\to} A \\ c &\longmapsto \text{id}_c \end{aligned}$$

and

$$\begin{aligned} A \times_O A &\overset{\circ}{\to} A \\ \langle g, f \rangle &\mapsto g \circ f. \end{aligned}$$

category of T-algebras The category of T-algebras or T-*models* in $\mathcal{E}$ is the entire subcategory **Alg**$(T, \mathcal{E})$ of the functor category $(T, \mathcal{E})$; its objects are the finite, product-preserving functions.

Cayley numbers A number system, denoted $\mathbb{O}$, which is modeled on $\mathbb{R}^8$ (the eightfold product of the real numbers $\mathbb{R}$ with itself).

chain If S is a partially ordered set then a chain in S is a subset $C \subseteq S$ that is linearly ordered (i.e., any two elements are comparable).

Change of Variables An axiom of the predicate calculus.

characteristic function If S is any set of integers, we let the *characteristic function* of S be

$$\chi_S(x) = \begin{cases} 1 \text{ if } x \in S \\ 0 \text{ if } x \notin S. \end{cases}$$

Church's theorem If T is a consistent extension of the Peano theory of the natural numbers, then T is undecidable.

Church's thesis The collection of general recursive functions exhausts the class of "effectively computable" functions.

class A collection of objects, formed according to strict rules, that may be larger than a set.

codomain See also *range*.

codomain of an arrow An operator assigns to each arrow f an object $b = \operatorname{cod} f$.

commutative A group G is said to be commutative if $g \cdot h = h \cdot g$ for any elements $g, h \in G$. There are similar definitions in other contexts. A commutative group is also called *abelian*.

commutative diagram A diagram is commutative if, for each pair of vertices c and c', any two paths formed from directed edges leading from c to c' yield, by composition of labels, equal arrows from c to c'.

compactness theorem In first-order logic, if every finite subset of a system S has a model, then S has a model.

complement Let X be the universal set and $A \subset X$. The complement of A, denoted ${}^cA$, is the set of those elements of X that are not in A. Sometimes called "the complement of A in X."

complete A system is complete if every true statement in the system is provable in that system.

complete induction See *strong mathematical induction.*

complete mathematical induction A method of induction whereby one proves $P(n)$ by assuming $P(1), \ldots, P(n-1)$. Also called *strong mathematical induction.*

complex numbers The number system, denoted by $\mathbb{C}$, that is modeled on $\mathbb{R} \times \mathbb{R}$ and contains the number i, which is a square root of -1.

complexity theory A theory that studies how many steps (as a function of n) it takes to solve a problem with n parameters or n pieces of data.

composition (in a category) The operation that assigns to each pair $\langle g, f \rangle$ of arrows with $\operatorname{dom} g = \operatorname{cod} f$ an arrow $g \circ f$ called their composite. Note that $g \circ f : \operatorname{dom} f \to \operatorname{cod} g$.

compound sentence In sentential logic, a sentence that is made up of elementary statements and connectives.

conjunction The conjunction of statements A and B is $A \wedge B$.

connective A device for joining together elementary statements. The standard connectives are $\sim$, $\vee$, $\wedge$, $\Rightarrow$, $\iff$. In some contexts, there are additional connectives called "nor" and "nand," denoted by $\downarrow$ and $|$.

consequence Suppose that R is a collection of relations on A. A relation on A is a *consequence* of R if it holds in every factor group of $[A]$ in which all the relations of R hold. The set of consequences of R is denoted by $\mathcal{C}(R)$.

conservative A system S is said to be conservative over a subsystem T if any result that can be proved in S can also be proved in T.

consistent A collection S of statements is *consistent* if the statement $A \wedge \sim A$ cannot be derived from S for any A.

constructible sets A model of set theory, consisting in effect of all those sets that can be described using first-order language using transfinite induction, that was used by Gödel to prove the consistency of the Axiom of Choice with the other axioms of set theory.

containment See also *subset.*

continuum The real number system, or any set with the same cardinality.

continuum hypothesis The proposition that the cardinality of the continuum is the very next cardinality after the cardinality of the natural numbers (i.e., that there are no cardinalities in between).

contradiction A statement that is always false, no matter what truth values are assigned to the component elementary statements.

contrapositive If $\mathbf{A} \Rightarrow \mathbf{B}$ is an implication, then $\sim \mathbf{B} \Rightarrow \sim \mathbf{A}$ is its contrapositive. The contrapositive is logically equivalent to the original implication.

converse If $\mathbf{A} \Rightarrow \mathbf{B}$ is an implication, then $\mathbf{B} \Rightarrow \mathbf{A}$ is its converse. The converse is logically independent of the original implication.

Cook's theorem The Satisfiability Problem is **NP**-complete.

coordinated We call a model M for a set S of statements *coordinated* if either of the following assertions holds:

1. If S is finite, then M is at most countable.

2. If S is infinite, then the cardinality of M does not exceed the cardinality of S.

corollary A corollary is a result that is an immediate consequence of some theorem.

coset The equivalence class obtained when we quotient a group by a normal subgroup.

Countable Axiom of Choice The Countable Axiom of Choice asserts that every countable set has a choice function.

countable chain condition The partially ordered set $(P, \leq)$ is said to satisfy the countable chain condition (ccc) if every pairwise incompatible (i.e., no two elements can be compared using $\leq$) subset is countable.

countable set A set with the same cardinality as the set of natural numbers.

counting argument A proof that depends on enumeration.

Craig's interpolation theorem Let A and B be first-order formulas such that $\models A \Rightarrow B$. Then there is a formula C such that $L(C) \Rightarrow [L(A) \cap L(B)]$ and such that $\models A \Rightarrow C$ and $C \Rightarrow B$.

crisp fuzzy set See *nonambiguous fuzzy set.*

cut A line in a proof of the form

$$[C \Rightarrow (A \vee B)] \wedge [(B \wedge C) \Rightarrow A] \Rightarrow [C \Rightarrow A].$$

cut elimination In proof theory, a process by which certain ambiguities (called *cuts*) can be eliminated from the formal structure of a proof. This method is useful in the search for propositional proofs.

decidable A system S is *decidable* if, for any statement A in S, either A or $\sim A$ can be proved in S.

decision problem The decision problem is to determine whether, given any statement A in Z_2, there is an effective method that yields a proof of either A or $\sim A$.

Dedekind completeness of the real numbers Let A and B be subsets of $\mathbb{R}$ such that

(i) $A \neq \emptyset$ and $B \neq \emptyset$;

(ii) $A \cup B = \mathbb{R}$;

(iii) $a \in A$ and $b \in B$ imply that $a < b$.

Then there exists exactly one element $x \in \mathbb{R}$ such that

(iv) If $u \in \mathbb{R}$ and $u < x$, then $u \in A$;

(v) If $v \in \mathbb{R}$ and $x < v$, then $v \in B$.

Dedekind cut The sets A and B in the definition of Dedekind completeness of the real numbers.

definable sets A model for set theory that can be used to prove that the Axiom of Choice is consistent with the other axioms of set theory. See *constructible sets.*

definition A statement that gives the meaning of a new term. Definitions are formulated in terms of other definitions and in terms of undefinables.

degree of recursive unsolvability Define a relation on functions by saying that $F \sim G$ if both F is recursive in G and G is recursive in F. This is in fact an equivalence relation. The resulting equivalence classes are called *degrees of recursive unsolvability.*

De Morgan's Laws for Sets If A, B, C are sets, then De Morgan's Laws for sets are

$$^c(A \cup B) = {}^cA \cap {}^cB$$

and

$$^c(A \cap B) = {}^cA \cup {}^cB.$$

deontic logic A variant of modal logic.

derivable Let S be a collection of statements. The statement A is *derivable* from S if, for some $B_1, \ldots, B_n \in S$, the statement $[B_1 \wedge \cdots \wedge B_n] \Rightarrow A$ is valid.

descriptive set theory Descriptive set theory endeavors to find languages that describe, or characterize, complexity classes.

deterministic Turing machine See *Turing machine.*

diagram A mathematical figure, composed of arrows and labels (which denote objects and arrows), that depicts relationships among objects in a category.

diagram scheme See *graph.*

difference of fuzzy sets The *difference* of fuzzy sets A and B is the fuzzy set AB defined by

$$(A - B)(x) = \max\{A(x) - B(x), 0\} \quad \forall x \in X.$$

Diophantine equation A polynomial equation, with integer coefficients, for which we seek integer solutions.

direct proof A method of proof that consists of moving in an efficient manner from the hypotheses to the desired conclusion.

Dirichletscher Schubfachschluss See *pigeonhole principle.*

discrete category A category is discrete if every arrow is an identity.

disjunction If **A** and **B** are statements, then their disjunction is $\mathbf{A} \vee \mathbf{B}$.

distributive Operations of addition and multiplication are said to be distributive if $a \cdot (b + c) = a \cdot b + a \cdot c$.

divisible group An abelian group G is said to be divisible if

$$\forall n \geq 1, \forall x, \exists y, [ny = x].$$

domain If f is a function from S to T, then we call S the domain of the function and denote it by $\operatorname{Dom} f$. There is a similar definition for a relation f.

Domain of an arrow An operator that assigns to each arrow f an object $a = \operatorname{dom} f$.

doxastic logic A variant of modal logic.

effectively computable function A function is effectively computable if

1. The procedure **m** is finite in length and time.

2. The procedure **m** is fully explicit and nonambiguous.

3. The procedure **m** is faultless and infallible.

4. The procedure **m** can be carried out by a machine.

element of We say that x is an element of the set S if x is one of the objects that composes S. We write $x \in S$.

elementary statement See *atomic statement.*

empty set The set with no elements.

equivalence class An *equivalence relation* partitions the given set S into disjoint subsets that are called equivalence classes.

equivalence relation Let S be a set and let $\sim$ be a relation on S. If $\sim$ satisfies

[**Reflexive**] $s \sim s$ for every $s \in S$,

[**Symmetric**] If $s \sim t$, then $t \sim s$,

[**Transitive**] If $s \sim t$ and $t \sim u$, then $s \sim u$,

then we say that $\sim$ is an *equivalence relation* on S.

Euclidean geometry The version of classical geometry in which the parallel postulate is taken to be true.

exponential complexity A problem is exponentially complex if there is a function of the form $e(n) = C \cdot 2^{k \cdot n}$ such that, with n pieces of data, the problem can be solved in at most $e(n)$ steps.

extension of a function Let $f : X \to Y$ be a function. Let $\widehat{X} \supset X$. We say that $\widehat{f} : \widehat{X} \to Y$ is an extension of f if the restriction of $\widehat{f}$ to X equals f.

factor group See *quotient group.*

Fagin's theorem The class **NP** is just the same as the class of generalized spectra.

false One of the two standard truth values (See *true*).

field A ring in which every nonzero element has a multiplicative inverse.

filter A filter over a nonempty set I is a set $D \subseteq \mathcal{P}(I)$ such that

1. $\emptyset \notin D$, $I \in D$;
2. If $X, Y \in D$, then $X \cap Y \in D$;
3. If $X \in D$ and $X \subseteq Y \subseteq I$, then $Y \in D$.

filtration A sequence of "worlds" (i.e., pairs of truth-functional assignments and sets of propositions) that sort out the formulas on which we wish to focus.

finite character A family of sets $\mathcal{F}$ has finite character if, for each set X, X belongs to $\mathcal{F}$ if and only if every finite subset of $\mathcal{F}$ belongs to $\mathcal{F}$.

finitely presented group See *group with generators and defining relations.* We say that a group is *finitely presented* if it is isomorphic to a group $[A; R]$ and both A and R are finite sets.

finite model theory A modal logic **L** has the *finite model property* if, for each formula X that is not valid in **L**, there is a finite model in which X is false.

finite type The finite types of a universe $\mathcal{U}$ over sets $A_0, \ldots, A_n$ are the sets generated from $A_0, \ldots A_n$ by closing under the operations

$$A, B \mapsto A \times B$$

$$(A \to B)_{\mathcal{U}}$$

$$\mathcal{P}(A).$$

first fundamental isomorphism of algebra If $\phi : G \to H$ is a surjective homomorphism of groups, then $H \cong G/\ker\phi$.

first incompleteness theorem Let T be a formal theory containing arithmetic. Then there is a sentence φ that asserts its own unprovability and is such that

(i) If T is consistent, then T does not prove φ.

(ii) If T is ω-consistent, then T does not prove $\sim \varphi$.

first-order language A *first-order language* has the following components:

- as variables,

$$x, y, z, w, z', y', z', w', x,'' y,'' z,'' w,'' \ldots;$$

- for each n, the n-ary function symbols and the n-ary predicate symbols;
- the symbols $\sim$, $\vee$, and $\exists$.

first-order logic We say that a first-order logic consists of the connectives $\wedge$, $\vee$, $\sim$, $\Rightarrow$, $\Longleftrightarrow$, the equality symbol $=$, and the quantifiers $\forall$ and $\exists$, together with an infinite string of variables $x, y, z, \ldots, x', y', z', \ldots$ and, finally, parentheses (,) to keep things readable.

first-order theory A *first-order theory* T has the following properties:

1. The language of T is a first-order language.
2. The axioms of T are the logical axioms and certain further nonlogical axioms.
3. The rules of inference in T are:

   Expansion Rule Infer $\mathbf{B} \vee \mathbf{A}$ from $\mathbf{A}$.

Contraction Rule Infer $\mathbf{A}$ from $\mathbf{A} \vee \mathbf{A}$.

Associative Rule Infer $(\mathbf{A} \vee \mathbf{B}) \vee \mathbf{C}$ from $\mathbf{A} \vee (\mathbf{B} \vee \mathbf{C})$.

Cut Rule Infer $\mathbf{B} \vee \mathbf{C}$ from $\mathbf{A} \vee B$ and $\sim \mathbf{A} \vee \mathbf{C}$.

$\exists$-Introduction Rule If x is not free in $\mathbf{B}$, infer $\exists x\mathbf{A} \to \mathbf{B}$.

for all The universal quantifier $\forall$.

forcing The method used by Paul J. Cohen to prove that the continuum hypothesis is independent of the Axiom of Choice.

formal system A finite collection of symbols and precise rules for manipulating these symbols to form certain combinations called "theorems." The rules should be quite explicit, and require no infinite processes to check.

Foundation Axiom See *Axiom of Regularity.*

Franks's theorem A result giving important functional equations for t-norms and t-conorms.

free and bound variables in the Lambda-calculus Let x be a variable in a λ-term M. We say that x is a *bound variable* if it occurs within a part of M having the form $\lambda x \bullet N$. Otherwise we say that the occurrence of x is free.

free group The free group $[A]$ on a set A is a group including A that has the following property: Every element of $[A]$ can be uniquely written in the form

$$a_1^{\pm 1} \cdot a_2^{\pm 1} \cdots a_k^{\pm 1},$$

where the a_i are in A and and no a_j appears adjacent to a_j^{-1}.

free product See *free product with amalgamation.* In the special case that ϕ is an isomorphism of the zero subgroups, we write $G * G'$ for $G *_\phi G'$ and we call $G * G'$ the *free product* of G and G'.

free product with amalgam H Let G and G' be any subgroups of a group L. Let $H = G \cap G'$. Let ϕ be the identity mapping from H to H. Then, by the theory developed above, there is a unique homomorphism from $G *_\phi G'$ to L that is the identity on G and G'. The image of this homomorphism is $\{G, G'\}$. In case this homomorphism is injective, we identify $G *_\phi G'$ with $\{G, G'\}$ and we say that $\{G, G'\}$ is the *free product* of G and G' with the amalgam H.

free product with amalgamation Let G and G' be groups. Let ϕ be an isomorphism of a subgroup H of G with a subgroup H' of G'. The free product of the groups G and G' with the amalgamation ϕ is the group

$$G *_\phi G' \equiv [G, G'; h = \phi(h)].$$

free subgroup A subset A of a group G is *free* if the homomorphism from $[A]$ to G that is the identity on A is injective.

free variable A variable of a function that satisfies one of the following stipulations:

(i) Every variable occurring in a formula of the form $x = y$, $x = c$, $c = c'$ or $R(t_1, \ldots, t_n)$ is free.

(ii) Variables in $\sim \mathbf{A}$, $\mathbf{A} \wedge \mathbf{B}$, $\mathbf{A} \vee \mathbf{B}$, $\mathbf{A} \Rightarrow \mathbf{B}$, and $\mathbf{A} \Longleftrightarrow \mathbf{B}$ are free if they are free in $\mathbf{A}$ or $\mathbf{B}$ separately.

(iii) The free occurrences of a variable in a formula of the form $\exists x, \mathbf{A}$ or $\forall x, \mathbf{A}$ are the same as the free occurrences of the variable itself, *except* that every occurrence of X is now considered bound.

function A function from a set S to a set T is a relation on S and T such that **(i)** if $s \in S$, then there is a $t \in T$ with (s, t) in the relation and **(ii)** if (s, t_1) and (s, t_2) are both in the relation, then $t_1 = t_2$.

functional A function that takes scalar (i.e., numerical) values.

functor A functor is a morphism (i.e., an arrow) of categories.

fuzzy logic A logic designed for the study of fuzzy sets and related ideas.

fuzzy point A fuzzy set A with the property that $A(x) > 0$ for only one point $x \in X$.

fuzzy set Let X be a universal set. A fuzzy set is a function $A : X \to [0, 1]$.

fuzzy set theory A version of set theory that discards the law of the excluded middle. It is a formalism for studying imprecision.

generalized continuum hypothesis This is the assertion that $2^{\aleph_j} = \aleph_{j+1}$.

general recursive function A function $f(x_1, \ldots, x_k)$ is said to be *general recursive* (or, sometimes, just "recursive") if there is a finite set of equations (expressed in terms of f) such that, for any choice of the numerals $n_1, \ldots, n_k$, there is a unique m such that $f(n_1, \ldots, n_k)$ can be deduced.

generator in a group The generators of a group have the property that their inverses and products give all the elements of the group.

Gödel completeness theorem Let S be a collection of statements in the propositional calculus that is consistent (i.e., it has no internal contradictions). Then there is a model for S.

Gödel incompleteness theorem Let Z_1 be the standard Peano formulation of arithmetic. Then there is a statement A in Z_1 such that neither A nor $\sim A$ can be proved from the axioms of Z_1.

Gödel number A number that is assigned to each wff in a formal system in order to prove the incompleteness theorem.

graph A graph is a set O of objects, a set A of arrows, and two functions

$$A \overset{\text{dom}}{\to} O$$

and

$$O \overset{\text{cod}}{\to} A.$$

In the graph, the set of composable pairs of arrows is the set

$$A \times_O A \equiv \{\langle g, f \rangle : g, f \in A \text{ and } \operatorname{dom} g = \operatorname{cod} f\}.$$

This set is called the *product over* O. See *diagram scheme.*

greater cardinality A set T has greater cardinality than a set S if there is a one-to-one function $f : S \to T$ but no one-to-one function $g : T \to S$.

ground type The most basic type on which finite type universes are built.

group A group is a set G that is equipped with a binary operation, which we think of as multiplication and denote by $\cdot$. If $g, h \in G$, then we assume that $g \cdot h \in G$ (this is the *closure property* of the group operation). The axioms for a group are:

(1) The group operation is associative: If $g, h, k \in G$, then $g \cdot (h \cdot k) = (g \cdot h) \cdot k$.

(2) There is an element $e \in G$ such that $e \cdot g = g \cdot e = e$ for every $g \in G$.

(3) For every element $g \in G$ there is an element $h \in$G such that $g \cdot h = h \cdot g = e$. We commonly denote this element h by g^{-1}.

group with generators and defining relations Suppose that R is a collection of relations on A. A relation on A is a consequence of R if it holds in every factor group of $[A]$ in which all the relations of R hold. The set of consequences of R is denoted by $\mathcal{C}(R)$. Let K_R be the normal subgroup generated by all expressions of the form $X \cdot Y^{-1}$ for any X, Y with $X = Y$ in R. The factor group $[A]/K_R$ is precisely the one in which the relations that hold are the consequences of R. We call $[A]/K_R$ the group with generators A and defining relations R. We designate that group by $[A; R]$.

Hahn–Banach theorem Let p be a sublinear functional on V and let ϕ be a linear functional defined on a subspace $W \subseteq V$ such that $\phi(x) \leq p(x)$ for all $x \in W$. Then there is a linear functional ψ on V that extends ϕ, and so that $\psi(x) \leq p(x)$ for all $x \in V$.

Hausdorff maximality principle If $\mathcal{R}$ is a transitive relation on a set S, then there exists a maximal subset of S that is linearly ordered by $\mathcal{R}$.

Hilbert's thesis The hypothesis that all mathematics can be formulated in first-order logic.

homomorphism Let G and H be groups and $h : G \to H$ be a function. We say that h is a homomorphism if $h(x \cdot y) = h(x) \cdot h(y)$. There is a similar concept for rings, fields, and other algebraic constructs.

ideal Let R be a ring and $J \subseteq R$ a subset. If J is a group under the addition operation in R and if $x \in R$ and $j \in J$ imply both $x \cdot j \in R$ and $j \cdot x \in R$, then J is an ideal of R.

identity (in a category) The operation that assigns to each object a an arrow $\mathrm{id}_a = 1_a : a \to a$.

identically true A statement that is true no matter what the truth values of its component variable letters. A tautology is identically true.

identity element For addition, the identity is an element 0 such that $x + 0 = x$ for every x. For multiplication, the identity is an element 1 such that $1 \cdot x = x$ for every x.

if-and-only-if The connective that means logical equivalence, denoted by $\iff$. If **A** and **B** are either both true or both false, then the statement $\mathbf{A} \iff \mathbf{B}$ is true. Otherwise the statement is false.

if-then The connective that means implication, denoted by $\Rightarrow$. If **B** is false, then $\mathbf{A} \Rightarrow \mathbf{B}$ is true. If both **A** and **B** are true, then $\mathbf{A} \Rightarrow \mathbf{B}$ is true. Otherwise, the implication is false.

image If f is a function from S to T, then the image of f, denoted $\operatorname{Im} f$, is the set of $t \in T$ such that there exists an $s \in S$ with $f(s) = t$. There is a similar definition for a relation f.

implication A statement of the form $\mathbf{A} \Rightarrow \mathbf{B}$, where **A** and **B** are component statements.

inaccessible cardinal The Axiom of Inaccessible Cardinals states the following. There is an uncountable cardinal A such that:

1. If $B < A$, then A is not the sum (union) of B cardinals each of which is $< A$.

2. If $B < A$, then the cardinality of $\mathcal{P}(B)$ is also $< A$.

independent Let S be a set of statements and let A be another statement that is not in S. We say that A is independent of S if S does not imply A.

induction See *mathematical induction*.

inductive definition A definition that specifies a sequence of objects inductively.

inductive hypothesis In mathematical induction, this is the assumption $P(n-1)$ that one uses to prove $P(n)$.

infinitesimal A number x in the nonstandard real number system such that $x \neq 0$ and $-a < x < a$ for every positive a.

integers That number system, denoted by $\mathbb{Z}$, consisting of the positive whole numbers, the negative whole numbers, and zero.

interpolant An interpolant for A and B with $A \Rightarrow B$ valid is a third formula C such that $A \Rightarrow C$ and $C \Rightarrow B$ are both valid.

interpolation The process of finding interpolants.

interpretation An interpretation of a model consists of a map $c_\alpha \to \overline{c}_\alpha$ from the constant elements of S to elements of M and a second map $R_\beta \to \overline{R}_\beta$ from the relation elements of S to elements of M such that all the statements of S are true in M.

intersection of sets If S and T are sets, then their intersection, denoted $S \cap T$, is the set consisting of all elements that are in both S and T.

inverse element For addition, the element $-x$ such that $x+(-x)=0$ is called the (additive) inverse element to x. For multiplication, the element x^{-1} such that $x \cdot x^{-1} = 1$ is called the (multiplicative) inverse element to x.

irrational numbers Those real numbers that have infinite, nonrepeating decimal expansions. Equivalently, those real numbers that cannot be expressed as the quotient of two integers.

isomorphism A homomorphism that is both one-to-one and onto.

isomorphism of graphs Two graphs are isomorphic if there is a combinatorial mapping matching up vertices and edges.

K A formulation of modal logic named after Saul Kripke.

k-ary relation Like a relation, or a binary relation, but with k arguments.

Lambda-calculus The formal rules for the λ-calculus are:

λ-Abstraction Syntax

(1) All variables and constants are λ-terms. These entities are called *atoms*.

(2) If M and N are λ-terms, then $(M\ N)$ is also a λ-term. With reference to a term of the form $(M\ N)$, we say that $(M\ N)$ is an *application* of M to N.

(3) If M is a λ-term and x is a variable, then $(\lambda x \bullet M)$ is also a λ-term. In this circumstance, $(\lambda x \bullet M)$ is said to be an *abstraction*, just because we form a function out of a term (expression).

Lambda-function A concept of function that concentrates on what the function does (as opposed to its name). These functions have two possible syntaxes:

- λ (variable) $\bullet$ (function_body)

or

- λ (signature) | (constraint) $\bullet$ (function_body)

Lambda-term achieved by substitution The notation

$$E[x/M]$$

denotes the λ-term obtained by replacing all free occurrences of the variable x in the λ-term E by the λ-term M. The rules governing substitution are:

1. $y[y/M] \equiv M$
2. $x[y/M] \equiv x$ provided that $y \neq x$
3. $(E\ F)[y/M] \equiv ((E[y/M])\,(F[y/M]))$
4. $(\lambda\, y \bullet E)[y/M] \equiv \lambda\, y \bullet E$
5. $(\lambda\, x \bullet E)[y/M] \equiv \lambda x \bullet (E[y/M])$
   provided that:
   (a) $x \neq y$
   (b) either x is not free in M or y is not free in E.

law of the excluded middle Aristotle's dictum that a sensible statement must be either true or false.

lemma A result that is derived from the axioms by means of a proof, but that has no intrinsic interest. A lemma is used as a stepping stone to proving a theorem.

limit ordinal An ordinal a is a *limit ordinal* if $a \neq 0$ and a is not a successor.

linear functional Let V be a vector space over the field k and $\ell : V \to k$ a function. We say that ℓ is a linear functional if $\ell(av + bw) = a\ell(v) + b\ell(w)$ for every $a, b \in k$ and every $v, w \in V$.

linearly ordered A partially ordered set S is said to be totally ordered if any two elements are comparable.

linear ordering Let the set S be equipped with a partial ordering. If $X \subseteq S$ is a subset with the property that any two elements of X *can* be compared under the partial ordering, then we say that X is linearly ordered. See *total ordering*.

logical axiom A logical axiom is one that is an artifact of the logic we use, and is not particular to the specific language (such as Euclidean geometry) being considered.

logically equivalent Two statements are logically equivalent if they have the same truth table. In other words, they are equivalent if, for each truth assignment, they have the same truth values.

Löwenheim–Skolem theorem The Löwenheim–Skolem theorem says that every model (in first-order logic) for a collection T of constants and relations has an elementary submodel whose cardinality does not exceed the cardinality of T, or (in case T is finite) is countable.

Martin's Axiom An axiom for set theory that has been proposed as an alternative to the continuum hypothesis.

mathematical induction A method of mathematical proof that is based on the ordering of the natural numbers $\mathbb{N}$. One proves a proposition $\mathbf{P}$ by assuming $P(n-1)$ and using that hypothesis to prove $P(n)$.

maximum principle for classes If every nonempty nest that is a subset of a nonempty set x has the property that its union is an element of x, then x has a maximal element.

measurable cardinal If there is a set S equipped with a nontrivial, countably additive, finite, real-valued measure defined on *all* subsets of S and such that the measure of each singleton is zero, then we say that S is a measurable cardinal.

membership degree For a fuzzy set A, the value $A(x)$ for a particular x.

membership function If X is a universal set, then a function $A : X \to I$, where I is the closed unit interval, is called a membership function (in fuzzy set theory).

metacategory A metagraph with two additional operations:

- the operation of *identity*, which assigns to each object a an arrow $\mathrm{id}_a = 1_a : a \to a$;
- the operation of *composition*, which assigns to each pair $\langle g, f \rangle$ of arrows with $\mathrm{dom}\, g = \mathrm{cod}\, f$ an arrow $g \circ f$ called their *composite*. Note that $g \circ f : \mathrm{dom}\, f \to \mathrm{cod}\, g$.

metagraph A collection of objects $a, b, c, \ldots$, arrows $f, g, h, \ldots$, and two operations:

- The operation of *Domain*, which assigns to each arrow f an object $a = \operatorname{dom} f$;

- The operation of *Codomain*, which assigns to each arrow f an object $b = \operatorname{cod} f$.

minimal model A model M for S is said to be *minimal* if it has least possible cardinality.

modal logic A multivalued logic allowing statements that are "necessarily true" or "possibly true."

model A *model* for a collection S of statements is a set M together with an interpretation of some constant and relation symbols in S. This "interpretation" consists of a map $c_\alpha \to \overline{c}_\alpha$ from the constant elements of S to elements of M and a second map $R_\beta \to \overline{R}_\beta$ from the relation elements of S to elements of M such that all the statements of S are true in M.

modus ponendo ponens The rule of logic that says "If $\mathbf{A}$ and $\mathbf{A} \Rightarrow \mathbf{B}$, then $\mathbf{B}$." Also called *modus ponens.*

modus tollens The rule of logic that says "If $\sim \mathbf{B}$ and $\mathbf{A} \Rightarrow \mathbf{B}$, then $\sim \mathbf{A}$.

monoid A category with just one object.

morphism An arrow or mapping in a category.

multivalued logic A logic in which statements are not merely assigned one of two truth values ("true" or "false"), but for which there is a larger collection of possible truth values.

μ-operator If $f(y, x_1, \ldots, x_n)$ is any function, then let $\mu_y f(y, x_1, \ldots, x_n)$ denote the function $g(x_1, \ldots, x_n)$ defined by these rules:

- We set $g(x_1, \ldots, x_n) = 0$ if, for any $\overline{x}_1, \ldots, \overline{x}_n$ (where $\overline{x}_1, \ldots \overline{x}_n$ are the images of $x_1, \ldots, x_n$ in some model), we have the nonvanishing of f: $f(y, \overline{x}_1, \ldots, \overline{x}_n) \neq 0$ for every y.

- If $x_1, \ldots, x_n$ are such that $f(y, \overline{x}_1, \ldots, \overline{x}_n) = 0$ for some y, then we set $g(x_1, \ldots, x_n) = a$, where a is the *least* y such that we have the indentity $f(y, \overline{x}_1, \ldots, \overline{x}_n) = 0$.

nand A connective that is sometimes used in Boolean algebra. The statement "$\mathbf{A}$ nand $\mathbf{B}$" means $\sim (\mathbf{A} \wedge \mathbf{B})$. The symbol for "nand" is $|$.

n-ary function An *n-ary function* from a set A to a set B is a function

$$f : \underbrace{A \times A \times \cdots \times A}_{n \text{ times}} \to B.$$

n-ary predicate A subset of the n-fold product $A \times A \times \cdots \times A$.

natural numbers The number system, denoted by $\mathbb{N}$, composed of the positive, whole numbers $1, 2, 3, \ldots$.

natural transformation Let S, T be functors from category C to category B. Then a natural transformation $\tau : S \dot{\to} T$ is a function that assigns to each object $c \in C$ an arrow $\tau_c \equiv \tau c : S(c) \to T(c)$ of B so that every arrow $f : c \to c'$ in C yields a commutative diagram.

nest A class that is linearly ordered by inclusion.

Nielsen–Schreier theorem Every subgroup of a free group is a free group.

nonambiguous fuzzy set A fuzzy set A that takes only the values 0 and 1.

nondeterministic Turing machine A nondeterministic Turing machine (NDTM) is a Turing machine with an extra write-only head that generates an initial guess for the Turing machine to evaluate.

noneuclidean geometry The version of classical geometry in which the parallel postulate is taken to be false.

nonlogical axiom An axiom that is an artifact of the language being used.

nonstandard analysis A number system containing the real numbers as well as certain infinitesimal numbers and certain infinitely large numbers.

nonstandard real numbers A number system that contains the standard real numbers and also infinitesimal numbers and infinitary numbers.

nor A connective that is sometimes used in Boolean algebra. The statement "**A** nor **B**" means $\sim (\mathbf{A} \vee \mathbf{B})$. The symbol for "nor" is $\downarrow$.

normal form Let T consist of one element from each right coset of H in G other than H itself. And let T' be formed similarly from G' and H'. A word on $G \cup G'$ is said to be in *normal form* if it can be written as $ht_1t_2 \cdots t_n$, where $h \in H$ and $t_1, \ldots, t_n \in T \cup T'$ and $t_i \in T$ if and only if $t_{i+1} \in T'$, $1 \leq i < n$.

normal subgroup A normal subgroup H in a group G is one with the property that $g^{-1}hg \in H$ for every $h \in H$ and $g \in G$.

not The logical connective that denotes negation, written $\sim$. If **A** is true, then $\sim A$ is false, and if **A** is false, then $\sim \mathbf{A}$ is true.

Novikov's theorem There is a finitely presented group that has an unsolvable word problem.

NP-complete problem Let Π be a problem in **NP**. We say that Π is **NP**-complete if Π is polynomially equivalent to every other problem in **NP**.

NP-hard problem A type of problem that includes both decision problems and search problems, and is equivalent to **NP**-complete.

NPI problem Let **NPC** denote the **NP**-complete problems. Then $\mathbf{NPC} \subsetneq [\mathbf{NP} \setminus \mathbf{P}]$. Moreover, it is known that $\mathbf{NPC} \neq [\mathbf{NP} \setminus \mathbf{P}]$. We let

$$\mathbf{NPI} \equiv [\mathbf{NP} \setminus \mathbf{P}] \setminus \mathbf{NPC}.$$

object A mathematical entity on which we operate.

object function A function T that assigns to each object $c \in C$ an object $T(c)$ of B.

Occam's razor A principle of philosophy, originally promulgated by William of Occam, that dictates that one should use the smallest and least redundant set of axioms possible.

ω-consistent A formal system T is *ω-consistent* if there is no formula W with one free variable such that $W(n)$ is a theorem for every natural number n but there is some x such that $W(x)$ is not a theorem. The property of ω-consistency is stronger than the property of consistency.

one-to-one A function f from S to T is said to be one-to-one if the condition $f(s_1) = f(s_2)$ entails $s_1 = s_2$.

one-to-one correspondence See *set-theoretic isomorphism.*

onto A function f from S to T is said to be onto if, for every $t \in T$, there is an $s \in S$ such that $f(s) = t$.

or The logical connective that disjoins two component statements, denoted by $\vee$. If either **A** or **B** or both is true then $\mathbf{A} \vee \mathbf{B}$ is true. The statement $\mathbf{A} \vee \mathbf{B}$ is false only when both **A** and **B** are false.

order See *order relation.*

ordered pair If S and T are sets, then an ordered pair of elements of S and T is an object of the form (s, t), where $s \in S$ and $t \in T$. See *product of two sets.*

order relation A relation $\mathcal{R}$ on a set S is said to be an order relation (or an *ordering*) on S if

1. For all $x, y \in S$, one and only one of the relations $x\mathcal{R}y$, $y\mathcal{R}x$, $x = y$ holds.

2. If $x, y, z \in S$ and $x\mathcal{R}y$ and $y\mathcal{R}z$, then $x\mathcal{R}z$.

ordinal number An *ordinal* is a set a that is well-ordered by the relation $\in$ and which is transitive.

Pairing Axiom See the *Power Set Axiom.*

parallel arrows The symbol $\downarrow\downarrow$ denotes the category with two objects a and b and just two arrows $a \to b$ and $b \to a$ that are not the identity arrows. We call two such arrows parallel arrows.

parallel postulate The axiom of classical Euclidean planar geometry that asserts that through any point off a given line there exists a unique line parallel to that given line.

partial function In case a function f has as its domain only a (possibly proper) *subset* of X, then we say that f is a *partial function* from X to Y.

partial ordering Let $\sim$ be a relation on a set S. If $\sim$ is reflexive, antisymmetric ($a \sim b$ and $b \sim a$ implies $a = b$), and transitive, then it is called a partial ordering.

partially ordered set A set equipped with a partial ordering.

Peano theory of the natural numbers An axiomatic treatment of arithmetic due to G. Peano.

pigeonhole principle The idea, originally due to Dirichlet, that if $n+1$ objects are placed in n boxes, then one box must contain (at least) two objects. See *Dirichletscher Schubfachschluss.*

polynomial complexity A problem is polynomially complex if there is a polynomial $p(n)$ such that, with n pieces of data, the problem can be solved in at most $p(n)$ steps.

polynomially equivalent We say that two problems Π_1 and Π_2 are polynomially equivalent if there is a translation of the language in which Π_1 is expressed into the language in which Π_2 is expressed that is of polynomial complexity. That is to say, there is a polynomial q such that any statement about Π_1 with n characters can be translated into a statement about Π_2 with at most $q(n)$ characters and conversely.

postulate See *axiom.*

power set Let S be any set. Then its *power set* $\mathcal{P}(S)$ is the collection of all subsets of S.

Power Set Axiom An axiom of Zermelo–Fraenkel set theory.

power set functor The functor $\mathcal{P} : \mathbf{Set} \to \mathbf{Set}$, which assigns to each set X its usual power set.

predicate calculus The rules for forming valid statements using the standard connectives as well as the quantifiers $\forall$ and $\exists$.

prenex normal form A standard form for a statement in the predicate calculus in which all quantifiers occur at the beginning of the statement.

prime ideal Let R be a ring and J an ideal in R. We say that J is a prime ideal if whenever $a, b \in R$ and $a \cdot b \in J$, then either $a \in J$ or $b \in J$.

prime ideal theorem Every Boolean algebra has a prime ideal.

prime number A positive integer whose only divisors or factors are 1 and itself. The first few primes are 2, 3, 5, 7, 11, 13, 17, ...

primitive recursive function A function $f(n_1, \ldots, n_k)$ from $\mathbb{Z}$ to $\mathbb{Z}$ is called primitive recursive (p.r.) if it is constructed by means of the following rules:

1. $f \equiv c$ for some constant c is p.r.
2. $f(n_1, \ldots, n_k) = n_i$, some $1 \leq i \leq k$, is p.r.
3. $f(n) = n + 1$ is p.r.
4. If $f(n_1, \ldots, n_k)$ and $g_1, \ldots, g_k$ are p.r., then so is $f(g_1, \ldots, g_k)$.
5. If $f(0, n_2, \ldots, n_)$ is p.r. and if $g(m, n_1, \ldots, n_k)$ is p.r. and if we have $f(n_1, n_2, \ldots, n_k) = g(f(n, n_2, \ldots, n_k), n, n_2, \ldots, n_k)$, then f is p.r.

Principle of Dependent Choices If R is a relation on a nonempty set S such that for every $x \in S$ there exists a $y \in S$ with xRy, then there is a sequence $\{x_j\}_{j=0}^{\infty}$ of elements of S such that

$$x_0 R x_1, \ x_1 R x_2, \ \ldots, x_n R x_{n+1}, \ldots$$

problem of class NP We say that a problem is of class **NP** if there is a polynomial p and a nondeterministic Turing machine with the property that, for any instance of the problem, the Turing machine will generate a guess of length n and come to a halt, generating an answer of "yes" or "no" (i.e., that this guess *is* a solution or *is not* a solution), in at most $p(n)$ steps.

problem of class P A problem is said to be of class **P** if there is a polynomial p and a (deterministic) Turing machine (DTM) for which every input of length n comes to a halt, with a yes/no answer, after at most $p(n)$ steps.

product (in category theory) In a graph, the set of composable pairs of arrows (called the product) is the set

$$A \times_O A \equiv \{\langle g, f \rangle : g, f \in A \text{ and } \operatorname{dom} g = \operatorname{cod} f\}.$$

product of fuzzy sets The product of two fuzzy sets A and B is the fuzzy set $A \cdot B$.

product of two sets If S and T are sets, then the product $S \times T$ of S and T is the set of all ordered pairs (s, t) with $s \in S$ and $t \in T$.

product of n sets If $S_1, \ldots, S_n$ are sets, then the product $S_1 \times \cdots S_n$ is the set of all n-tuples $(s_1, \ldots, s_n)$ with $s_j \in S_j$, $j = 1, \ldots, n$.

product over O In a graph (in category theory), the set of composable pairs of arrows given by

$$A \times_O A \equiv \{\langle g, f \rangle : g, f \in A \text{ and } \operatorname{dom} g = \operatorname{cod} f\}.$$

proof A sequence of statements that lead from the definitions and the axioms to some statement that we wish to establish. The only allowable rule for moving from the nth statement to the $(n+1)$st statement is to apply *modus ponens* (or, equivalently, *modus tollens*).

proof by contradiction A method of proof that proceeds by denying the desired conclusion and then arriving at a contradiction.

proof by enumeration A proof that consists of a (often sophisticated) counting argument.

proof by induction A proof that employs a mathematical induction argument.

proper subset We say that the set A is a proper subset of the set B, and we write $A \subset B$, if $A \subseteq B$ but $A \neq B$. See *subset.*

proposition Like a theorem, but of lesser importance.

propositional calculus The rules for forming valid statements using the standard connectives.

propositional function A combination of variable letters using the standard connectives.

pspace The class of languages that can be accepted by a Turing machine in polynomial space.

Pythagorean theorem Let $\triangle$ be a right triangle with legs a, b and hypotenuse c. Then $c^2 = a^2 + b^2$.

quaternions A number system, denoted $\mathbb{H}$, modeled on $\mathbb{R} \times \mathbb{R} \times \mathbb{R} \times \mathbb{R}$.

quotient group Let G be a group and H a normal subgroup. The quotient group of G by H is the collection of cosets of H in G.

Ramsey theory A branch of mathematics that involves sophisticated counting arguments and the classification of sets into subsets.

range If f is a function from S to T, then we call T the range of f and denote it by $\operatorname{Ran} f$. There is a similar definition for a relation f. Also called the *codomain.*

rational numbers The number system, denoted by $\mathbb{Q}$, composed of all quotients of integers.

real numbers A number system, denoted by $\mathbb{R}$, that is the completion of the rational numbers. The collection of all rational numbers and irrational numbers.

recursive function See *general recursive function.*

recursive in the function G A function F is recursive in the function G if F may be built from G using the steps allowed in inductively forming a recursion.

recursive set A set S is said to be *recursive* if its characteristic function χ_S is general recursive.

recursively enumerable A set S is *recursively enumerable* if S is either empty or is the image of a general recursive function.

recursively presented group A group is *recursively presented* if it is isomorphic to a group $[A; R]$, where A is finite and R is recursively enumerable. See *group with generators and defining relations.*

recursivity of one function in another A function F is recursive in the function G if F may be built from G using the steps allowed in inductively forming a recursion.

reduction The process of cancelling juxtaposed inverses in a formal element of a free group.

reflexive A relation $\mathcal{R}$ is reflexive if $x\mathcal{R}x$ for every x in the domain.

relation A relation on two sets S and T is a subset of $S \times T$.

relation as a consequence of other relations A relation on A is a *consequence* of R if it holds in every factor group of $[A]$ in which all the relations of R hold.

relation holding in a factor group A relation is said to hold in a factor group $[A]/K$ if X and Y (See *relation in a group*) lie in the same coset of K, equivalently, if $XY^{-1} \in K$.

relation in a group A relation on a set $A \subset G$ is an expression of the form $X = Y$, where X and Y are words on A.

relation symbol A symbol denoting a k-ary relation.

relative consistency In view of Gödel's second incompleteness theorem, this is the notion of proving consistency of a theory in the context of consistency of some other theory.

resolution rule If C and D are clauses (i.e., disjunctions of propositional variables or their negations) and if $x \in C$ and $\sim x \in D$, then the resolution rule applied to C and D is the inference $C \wedge D$ implies $(C \setminus \{x\} \cup (D \setminus \{\sim x\})$.

restriction If $\phi : G \to H$ is a mapping or a homomorphism and if $B \subset G$ is a subset, then we let $\phi\big|_B$ denote the restriction of ϕ to B; this means that we just consider B to be the domain of ϕ.

ring An algebraic system equipped with operations of addition and multiplication satisfying a set of standardized axioms.

Rule of Equality An axiom of the predicate calculus.

Rule of Inference—*Modus Ponens* An axiom of the propositional calculus.

Rule of the propositional calculus An axiom of the propositional calculus.

Rule of Specialization An axiom of the predicate calculus.

Russell's paradox A paradox produced by Bertrand Russell that necessitates certain restrictions on the way that we form sets.

same equivalence class Sets S and T have the same equivalence class if the two sets are set-theoretically isomorphic.

Satisfiability Problem Let $\mathcal{P}$ be a finite collection of statements from the propositional calculus. That is, it is a collection of sentences formed from finitely many atomic sentences $A_1, \ldots, A_k$ using the basic connectives. The Satisfiability Problem is this: Is there an assignment of truth values to $A_1, \ldots, A_k$ so that all the statements in $\mathcal{P}$ are true?

satisfiable See *consistent.*

Schreier's theorem See *free product with amalgamation* and *normal form.* Schreier's theorem states that every coset in $G *_\phi G'$ contains exactly one word in normal form.

Schroeder–Bernstein theorem Let S and T be sets. If there exists a one-to-one function $f : S \to T$ and a one-to-one function $g : T \to S$, then S and T have the same cardinality.

second incompleteness theorem Let T be a consistent formal theory containing arithmetic. Then

$$T \text{ does not prove } \operatorname{Consis} T.$$

Here $\operatorname{Consis} T$ is the sentence asserting that T is consistent.

second-order logic Second-order logic allows us to quantify over subsets of M (the model) and functions F mapping $M \times M$ into M. Compare *first-order logic.*

sequence A *sequence* on a set S is a function $\varphi : \mathbb{N} \to S$. We often write $s_1, s_2, \ldots$ or $\{s_j\}_{j=1}^{\infty}$.

set A collection of objects.

set-builder notation The notation, such as

$$T = \{m \in \mathbb{Z} : m^2 - 3m + 5 > 0\},$$

that we use to describe a set by specifying (i) a set of which it is a subset and (ii) a condition that describes the set.

set formulation of induction See *complete mathematical induction.*

set-theoretic difference If S and T are subsets of a given set X, then the set-theoretic difference of S and T, denoted $S \setminus T$, is the set of elements of S that do not lie in T.

set-theoretic isomorphism A function $f : S \to T$ is called a set-theoretic isomorphism if it is both one-to-one and onto.

solvable A question is solvable if its associated decision problem is solvable.

Souslin tree An uncountable tree in which every chain and every antichain is countable.

spectrum The spectrum of a first-order sentence is the set of cardinalities of its finite models.

Stirling's formula An asymptotic formula for the value of $n!$ (n factorial).

strong mathematical induction See *complete induction.* The inductive method in which the nth statement is proved by assuming that statements $1, 2, \ldots, (n-1)$ are all true.

subgraph problem Let G be a graph. Let H be another graph. The question is whether G contains a subgraph that is isomorphic to H.

subgroup A subset $H \subseteq G$ is called a subgroup if (i) $e \in H$, (ii) whenever $g, h \in H$, then $g \cdot h \in H$, and (iii) whenever $g \in H$, then $g^{-1} \in H$.

subgroup generated by B If $B \subseteq G$ is any set (not necessarily a subgroup), then we let $\{B\}$ denote the *subgroup* generated by B. Thus $\{B\}$ consists of all (finite) expressions of the form

$$b_1^{\pm 1} \cdot b_2^{\pm 1} \cdots b_k^{\pm 1}$$

formed with elements $b_j \in B$.

subset The set A is a subset of the set B, written $A \subseteq B$ or $A \subset B$, if $x \in A$ implies $x \in B$.

successor We say that an ordinal a is a *successor* if there is an ordinal b such that $b = a + 1$.

Sum Axiom An axiom of Zermelo–Fraenkel set theory.

superset The set B is a superset of the set A precisely when A is a subset of the set B.

surjective See *onto.*

symmetric A relation $\mathcal{R}$ is symmetric if (for all x, y in the domain) $x\mathcal{R}y$ whenever $y\mathcal{R}x$.

symmetric difference Let S, T be subsets of a given set X. The set-theoretic difference of S and T, denoted $S \triangle T$, is $[S \setminus T] \cup [T \setminus S]$.

Tarski's cardinality theorem Let $\mathbf{m}$ be an infinite cardinal. Let T be an $\mathbf{m}$-theory (i.e., a theory whose language has a set of nonlogical symbols having cardinality less than or equal to $\mathbf{m}$) having an infinite model. Then T has a model of cardinality $\mathbf{m}$.

tautologically implies Let S be a set of statements and φ a statement. We say that S tautologically implies φ if any model that makes all the statements of S true also makes φ true. We write $S \models \varphi$.

tautology A statement that is always true, no matter what truth values are assigned to the component elementary statements.

***t*-conorm** See *triangular conorm.*

temporal logic A variant of modal logic.

term formation rule If $t_1, \ldots, t_n$ are terms and if f is an n-ary function symbol of the language, then the expression $f(t_1, \ldots, t_n)$ is a term.

theorem A statement, of some significance, that can be proved from the axioms (and, implicitly, from other theorems, propositions, and lemmas).

there exists The existential quantifier $\exists$.

third-order logic Augments second-order logic by treating sets of function and more abstract constructs.

***t*-norm** See *triangular norm.*

total ordering See *linear ordering.*

transfinite induction A generalized induction process that is modeled on the ordinal numbers.

transitive We say that a set a is transitive if $b \in a$ and $c \in b$ implies $c \in a$.

transitive A relation is transitive if (whenever x, y, z are in the domain) $x\mathcal{R}y$ and $y\mathcal{R}z$ implies $x\mathcal{R}z$.

traveling salesman problem The problem of finding the best route for a salesman who must visit several different cities.

triangular conorm If T is a triangular norm (or t-norm), then $S(a, b) = 1 - T(1 - a, 1 - b)$ is a triangular conorm, or t-conorm.

triangular norm A function $T : I \times I \to I$ is a triangular norm if

(i) $T(a, 1) = a \quad \forall a \in I$;

(ii) $T(a, b) \leq T(u, v)$ if $a \leq u$, $b \leq v$;

(iii) $T(a, b) = T(b, a)$;

(iv) $T(T(a, b), c) = T(a, T(b, c))$.

trichotomy of set theory For any two sets A and B, either there is an injection of A into B or there is an injection of B into A.

true One of the two standard truth values (See *false*).

truth table A tableau that is used to assess the truth value of a statement.

truth value Each statement is assigned a truth value. Most commonly, the truth values are "true" and "false." There are also multivalued logics.

Tukey–Tychanoff lemma Let $\mathcal{F}$ be a nonempty family of sets. If $\mathcal{F}$ has finite character, then $\mathcal{F}$ has a maximal element (with respect to the partial ordering $\subseteq$).

Turing machine A device for performing effectively computable operations. It will consist of a device through which an infinite paper tape is fed. The tape is divided into an infinite sequence of congruent boxes. Each box has either the numeral 0 or the numeral 1 in it. The Turing machine has finitely many "states" $S_1, S_2, \ldots, S_n$. In any given state of the Turing machine, one of the boxes is being scanned.

After scanning the designated box, the Turing machine does one of three things:

(1) It either erases the numeral 1 that appears in the scanned box and replaces it with a 0, or it erases the numeral 0 that appears in the scanned box and replaces it with a 1, or it leaves the box unchanged.

(2) It moves the tape one box (or one unit) to the left or to the right.

(3) It goes from its current state S_j into a new state S_k.

Also called a *deterministic Turing machine.*

Tychanoff's product theorem The product of (any number of) compact sets is compact in the product topology.

ultrafilter A filter D over I is called an *ultrafilter* if, for every $X \subseteq I$, either $X \in D$ or $I \setminus X \in D$.

uncountable set A set with cardinality greater than the cardinality of the natural numbers.

undecidable A system S is *undecidable* if there is a statement $\mathbf{A}$ for which neither a proof of $\mathbf{A}$ nor a proof of $\sim \mathbf{A}$ can be found.

undefinable A term that cannot be defined using other terms. The meaning of an undefinable should be self-evident.

Union Axiom See the *Sum Axiom.*

union of sets If S and T are sets, then their union, denoted $S \cup T$, is the set consisting of all elements that are either in S or in T or in both.

univalent See *one-to-one.*

universal algebra A set together with a collection of n-ary or infinitary operations.

universal formula A formula of the form $\forall x_1, \ldots \forall x_k \theta$, where θ is quantifier-free.

universal set The largest set in the discussion. That set of which all others are subsets.

universe A pair $\mathcal{U} = (\mathrm{Set}_{\mathcal{U}}, \mathrm{Fn}_{\mathcal{U}})$. Here, $\mathrm{Set}_{\mathcal{U}}$ is the sets of $\mathcal{U}$ and $\mathrm{Fn}_{\mathcal{U}}$ is the functions of $\mathcal{U}$.

unsolvable A question is unsolvable if its associated decision problem is unsolvable.

upper bound for a chain An element u is an upper bound of the chain C if $c \leq u$ for every $c \in C$.

valid A statement that is true in every model.

variable letters Symbols used to denote elementary statements, numbers, and other components of a logical statement.

Venn diagram A figure that uses regions in the plane to illustrate sets, their intersections, unions, and containments.

well-formed formula These are the "grammatically correct" formulas in a given language, formed according to the following rules:

1. The statements $x = y$, $x = c$, $c = c'$ are wffs, where x an dy are variables and c, c' are constants.

2. If R is a k-ary relation and if each of $t_1, \ldots, t_n$ is either a variable or a constant, then $R(t_1, \ldots, t_n)$ is a wff.

3. If $\mathbf{A}$ and $\mathbf{B}$ are wffs, then so are $\sim \mathbf{A}$, $\mathbf{A} \wedge \mathbf{B}$, $\mathbf{A} \vee \mathbf{B}$, $\mathbf{A} \Rightarrow \mathbf{B}$, and $\mathbf{A} \iff \mathbf{B}$.

4. If $\mathbf{A}$ is a wff, then so are $\exists x, \mathbf{A}$ and $\forall x, \mathbf{A}$.

well-ordering An order relation $<$ on S is said to *well-order* S if each $B \subseteq S$ has a least element: there exists a $b_0 \in B$ such that $b \not< b_0$ for all $b \in B$.

wff See *well-formed formula.*

word In the definition of *free group*, the expressions

$$a_1^{\pm 1} \cdot a_2^{\pm 1} \cdots a_k^{\pm 1}.$$

word problem The decision problem for $\mathcal{C}(R)$ (See *consequence*). The word problem for $[A; R]$ is solvable if and only if $\mathcal{C}(R)$ is recursive. If G is a finitely presented group, then G is isomorphic to a group $[A; R]$ with A finite. We say that the word problem for G is *solvable* or *unsolvable* according to whether the word problem for $[A; R]$ is solvable or unsolvable.

Zermelo–Fraenkel set theory One of the standard axiomatic treatments of set theory.

zero subgroup The subgroup of a group G consisting of $\{e\}$ alone.

Zorn's lemma Let S be a nonempty, partially ordered set with the property that every chain in S has an upper bound. Then S has a maximal element.

Guide to the Literature

Classical Tracts on Logic

[**BAC**] H. Bachmann, *Transfinite Zahlen*, Springer-Verlag, Berlin, 1955.

[**BAR**] J. Barwise, ed. *Handbook of Mathematical Logic*, North-Holland, Amsterdam, 1977.

[**BER**] P. Bernays, *Axiomatic Set Theory*, North-Holland, Amsterdam, 1958.

[**BET1**] E. Beth, *Foundations of Mathematics*, North-Holland, Amsterdam, 1959.

[**BET2**] E. Beth, *Formal Methods*, Reidel, Dordrecht, 1962.

[**BRO**] L. E. J. Brouwer, *Collected Works, Vol. 1—Philosophy and Foundations of Mathematics*, A. Heyting, ed., North-Holland, Amsterdam, 1975.

[**CAR1**] R. Carnap, *Abriss der Logistik mit Besonderer Berucksichtigung der Relationstheorie und Ihrer Anwendung*, Springer-Verlag, Vienna, 1929.

[**CAR2**] R. Carnap, *The Logical Syntax of Language*, K. Paul, Trench, Trubner, & Co., London, 1937.

[**CAR3**] R. Carnap, *Foundations of Logic and Mathematics*, University of Chicago Press, Chicago, 1939.

[**CAR4**] R. Carnap, *Introduction to Semantics*, Harvard University Press, Cambridge, 1942.

[**CAR5**] R. Carnap, *The Continuum of Inductive Methods*, University of Chicago Press, Chicago, 1952.

[CAR6] R. Carnap, *Formalization of Logic*, Harvard University Press, Cambridge, 1947.

[CAR7] R. Carnap, *Introduction to Symbolic Logic*, Dover, New York, 1958.

[CHU1] A. Church, *A Bibliography of Symbolic Logic*, Association for Symbolic Logic, Menaster, Wisconsin, 1937–1939.

[CHU2] A. Church, *The Calculi of Lambda-Conversion*, Princeton University Press, Princeton, 1941.

[CHU3] A. Church, *Introduction to Mathematical Logic*, Princeton University Press, Princeton, 1956.

[COH] P. J. Cohen, *Set Theory and the Continuum Hypothesis*, Benjamin, New York, 1966.

[CUR1] H. Curry, *A Theory of Formal Deducibility*, Notre Dame Mathematical Lectures, University of Notre Dame, 1950.

[CUR2] H. Curry, *Outlines of a Formalist Philosophy of Mathematics*, North-Holland, Amsterdam, 1951.

[CUR3] H. Curry, *Leçons de logique algebrique*, Gauthier-Villars, Paris-Louvain, 1952.

[CUR4] H. Curry, *Foundations of Mathematical Logic*, McGraw-Hill, New York, 1963.

[CUF] H. Curry and R. Feys, *Combinatory Logic*, North-Holland, Amsterdam, 1958.

[END] H. Enderton, *A Mathematical Introduction to Logic*, 2nd ed., Academic Press, New York, 2001.

[FRA] A. A. Fraenkel, *Abstract Set Theory*, North-Holland, Amsterdam, 1953.

[FRE] G. Frege, *Foundations of Arithmetic*, Blackwell, Oxford, 1959.

[GOD1] K. Gödel, *On Undecidable Propositions of Formal Mathematical Systems*, Princeton University Press, Princeton, 1934.

[GOD2] K. Gödel, *The Consistency of the Axiom of Choice and of the Generalized Continuum Hypothesis with the Axioms of Set Theory*, Princeton University Press, Princeton, 1940.

[HAL1] P. R. Halmos, *Naive Set Theory*, Van Nostrand, Princeton, 1960.

[HAL2] P. R. Halmos, *Algebraic Logic*, Chelsea, New York, 1962.

[HAS] G. Hasenjaeger and H. Scholz, *Grundzüge der Mathematischen Logik*, Springer-Verlag, Berlin, 1961.

[HER] H. Hermes, *Aufzählbarkeit, Entscheidbarkeit, Berechenbarkeit*, Springer-Verlag, Berlin, 1961.

[HEY] A. Heyting, *Intuitionism*, North-Holland, Amsterdam, 1956.

[HIA] D. Hilbert and W. Ackermann, *Principles of Mathematical Logic*, Chelsea, New York, 1950.

[HIB] D. Hilbert and P. Bernays, *Grundlagen der Mathematik*, Springer-Verlag, Berlin, Vol. I (1934), Vol. II (1939).

[HOH] F. Hohn, *Applied Boolean Algebra*, Macmillan, New York, 1960.

[HRJ] K. Hrbacek and T. Jech, *Introduction to Set Theory*, 3rd ed., Marcel Dekker, New York, 1999.

[JEC] T. J. Jech, *The Axiom of Choice*, North-Holland, Amsterdam, 1973.

[KAM] E. Kamke, *Theory of Sets*, Dover, New York, 1950.

[KML] W. Kamlah, *Logical Propaedutic*, University Press of America, Lanham, Maryland, 1984.

[KLE] S. C. Kleene, *Introduction to Metamathematics*, Elsevier, New York, 1974.

[KLV] S. C. Kleene and R. E. Vesley, *The Foundations of Intuitionistic Mathematics*, North-Holland, Amsterdam, 1965.

[LAD] J. Ladrière, *Les Limitations Internes des Formalismes*, E. Nauwelaerts, Paris, 1957.

[LOR] P. Lorenzen, *Einfürung in die Operative Logik und Mathematik*, Springer-Verlag, Berlin, 1955.

[LUX] W. A. J. Luxemburg, *Nonstandard Analysis*, California Institute of Technology, Pasadena, 1962.

[MEN] E. Mendelson, *Introduction to Mathematical Logic*, 4th ed., Chapman and Hall, London, 1997.

[MOS] A. Mostowski, *Sentences Undecidable in Formalized Arithmetic*, North-Holland, Amsterdam, 1952.

[PET] R. Péter, *Rekursive Funktionen*, Akademie-Verlag, Budapest, 1951.

[QUI1] W. V. Quine, *Methods of Logic*, Holt, New York, 1950.

[QUI2] W. V. Quine, *Mathematical Logic*, Harvard University Press, Cambridge, 1951.

[QUI3] W. V. Quine, *Algebraic Logic and Predicate Functors*, UCB, New York, 1971.

[ROB1] A. Robinson, *On the Metamathematics of Algebra*, North-Holland, Amsterdam, 1951.

[ROB2] A. Robinson, *Complete Theories*, North-Holland, Amsterdam, 1956.

[ROB3] A. Robinson, *Nonstandard Analysis*, North-Holland, Amsterdam, 1966.

[ROG] H. Rogers, Jr., *Theory of Recursive Functions and Effective Computability*, McGraw-Hill, New York, 1917.

[ROS] P. Rosenblum, *Elements of Mathematical Logic*, Dover, New York, 1950.

[RSR] J. B. Rosser, *Deux Esquisses de Logique*, Gauthier-Villars, Paris, 1955.

[RR1] H. Rubin and J. Rubin, *Equivalents of the Axiom of Choice*, North-Holland, Amsterdam, 1970.

[RR2] H. Rubin and J. Rubin, *Equivalents of the Axiom of Choice*, II, North-Holland, Amsterdam, 1985.

[SCH] J. R. Schoenfield, *Mathematical Logic*, Addison-Wesley, Reading, Massachusetts, 1967.

[SCT] K. Schütte, *Beweistheorie*, Springer, Berlin, 1960.

[SIE] W. Sierpinski, *Cardinal and Ordinal Numbers*, Panstworwe Wydawn, Warsaw, 1958.

[SIK] R. Sikorski, *Boolean Algebras*, Springer-Verlag, Berlin, 1960.

[SMU] R. Smullyan, *Theory of Formal Systems*, Princeton University Press, Princeton, 1961.

[SUP1] P. Suppes, *Introduction to Logic*, Van Nostrand, Princeton, 1962.

[SUP2] P. Suppes, *Axiomatic Set Theory*, Van Nostrand, Princeton, 1972.

[TAR1] A. Tarski, *Introduction to Logic and to the Methodology of Deductive Science*, Oxford University Press, New York, 1946.

[TAR2] A. Tarski, *A Decision Method for Elementary Algebra and Geometry*, Rand Corporation, Santa Monica, 1948.

[TAR3] A. Tarski, *Logic, Semantics, Metamathematics*, Clarendon, Oxford, 1956.

[TMR] A. Tarski, A. Mostowski, and R. Robinson, *Undecidable Theories*, North-Holland, Amsterdam, 1953.

[WEY] H. Weyl, *Das Kontinuum*, Chelsea, New York, 1932.

[WIL] R. Wilder, *Introduction to the Foundations of Mathematics*, John Wiley and Sons, New York, 1952.

Modern Treatments of Subjects in Logic

[ABE] O. Aberth, *Computable Analysis*, McGraw-Hill, New York, 1980.

[AGM] S. Abramsky, D. M. Gabbay, and T. S. E. Maibaum, *Handbook of Logic in Computer Science*, vols. 1–3, Clarendon Press, Oxford, 1992, 1994.

[ATA] K. T. Atanassov, *Intuitionistic Fuzzy Sets*, Physica-Verlag, Heidelberg, 1999.

[FEE] M. Beeson, *Foundations of Constructive Mathematics*, Springer-Verlag, Berlin, 1985.

[BIS] E. Bishop, *Foundations of Constructive Analysis*, McGraw-Hill, New York, 1967.

[BIB] E. Bishop and D. Bridges, *Constructive Analysis*, Springer-Verlag, Berlin, 1985.

[CHE] B. F. Chellas, *Modal Logic—An Introduction*, Cambridge University Press, Cambridge, 1980.

[CUT] N. Cutland, ed., *Nonstandard Analysis and Its Applications*, Cambridge University Press, Cambridge, 1988.

[DAV1] M. Davis, *Applied Nonstandard Analysis*, Wiley Interscience, New York, 1977.

[DAV2] M. Davis, *Computability and Unsolvability*, Dover, New York, 1982.

[DLJ] D. Dumitrescu, B. Lazzerini, L. C. Jain, *Fuzzy Sets and their Application to Clustering and Training*, CRC Press, Boca Raton, 2000.

[DUM] M. A. E. Dummett, *Elements of Intuitionism*, Clarendon Press, Oxford, 1977.

[BRI] Encyclopedia Britannica Web article on universal algebra,
`http://www.britannica.com/bcom/eb/article`
`/2/0,5716,120652+2+111000,00.html`.

[GHR] D. M. Gabbary, C. J. Hogger, and J. A. Robinson, *Handbook of Logic in Artificial Intelligence and Logic Programming*, vol. 1, Oxford Science Publications, Oxford, 1993.

[GAJ] M. R. Garey and D. S. Johnson, *Computers and Intractability: A Guide to the Theory of NP-Completeness*, W. H. Freeman and Co., San Francisco, 1991.

[HA2] S. Givant and P. R. Halmos, *Logic as Algebra*, Mathematical Association of America, Washington, D.C., 1998.

[HUC] G. E. Hughes and M. J. Cresswell, *An Introduction to Modal Logic*, Methuen, London, 1968.

[LIN] T. Lindstrøm, An invitation to nonstandard analysis, in *Nonstandard Analysis and Its Applications*, N. Cutland, ed., Cambridge University Press, Cambridge, 1988.

[LOW] P. A. Loeb and M. Wolff, eds., *Nonstandard Analysis for the Working Mathematician*, Kluwer, Dordrecht, 2000.

[NEL] E. Nelson, Internal set theory: a new approach to nonstandard analysis, *Bull. Am. Math. Soc.* 83(1977), 1165–1198.

[NIS] N. Nissanke, *Introductory Logic and Sets for Computer Scientists*, Addison-Wesley, Reading, 1999.

[PIA] M. Piastra, Probability, modal logic, and fuzzy logic,
`http://citeseer.nj.nec.com/piastra98probability.html`.

[POP] S. Popkorn, *First Steps in Modal Logic*, Cambridge University Press, Cambridge, 1994.

[PRA] D. Prawitz, *Natural Deduction: A Proof-Theoretical Study*, Almqvist & Wiksell, Amsterdam, 1965.

[RR1] H. Rubin and J. Rubin, *Equivalents of the Axiom of Choice*, North-Holland, Amsterdam, 1970.

[RR2] H. Rubin and J. Rubin, *Equivalents of the Axiom of Choice*, II, North-Holland, Amsterdam, 1985.

[SIM] A. K. Simpson, *The Proof Theory and Semantics of Intuitionistic Modal Logic*, Ph.D. thesis, University of Edinburgh, Department of Computer Science, 1994.

[SMU1] R. Smullyan, *Fundamentals of Logic*, Prentice-Hall, Englewood Cliffs, 1962.

[SMU2] R. Smullyan, *First Order Logic*, Springer-Verlag, Berlin, 1968.

[SMU3] R. Smullyan, *Forever Undecided*, Knopf, New York, 1987.

[SMU4] R. Smullyan, *Gödel's Incompleteness Theorem*, Oxford University Press, New York, 1992.

[SMU5] R. Smullyan, *Recursion Theory for Metamathematics*, Oxford University Press, New York, 1993.

[SMU6] R. Smullyan, *Diagonalization and Self-Reference*, Clarendon Press, Oxford, 1994.

[SMU7] R. Smullyan, *Set Theory and the Continuum Problem*, Clarendon Press, Oxford, 1996.

[STA] *Stanford Encyclopedia of Philosophy*, `http://plato.stanford.edu/entries/logic-modal/`.

[TRO1] A. S. Troelstra, *Principles of Intuitionism*, Lecture Notes in Mathematics 95, Springer-Verlag, Berlin, 1969.

[TRO2] A. S. Troelstra, *Metamathematical Investigation of Inituitionistic Arithmetic and Analysis*, Lecture Notes in Mathematics 344, Springer-Verlag, Berlin, 1973.

[VDAL] D. van Dalen, *Brouwer's Cambridge Lectures on Intuitionism*, Cambridge University Press, Cambridge, 1981.

References from Computer Science

[AHU] A. V. Aho, J. E. Hopcroft, and J. D. Ullman, *The Design and Analysis of Computer Algorithms*, Addison-Wesley, Reading, 1974.

[AHUL] A. V. Aho and J. D. Ullman, *The Theory of Parsing, Translation, and Compiling—Volume 1: Parsing*, Prentice-Hall, Englewood Cliffs, 1972.

[APO] K. R. Apt and E.-R. Olderog, *Verification of Sequential and Concurrent Programs*, Springer-Verlag, Berlin, 1991.

[BAC] R. C. Backhourse, *Program Construction and Verification*, Prentice-Hall, Englewood Cliffs, 1986.

[BEK] K. Broda, S. Eisenbach, H. Khoshnevisan, and S. Vickers, *Reasoned Programming*, Prentice-Hall, Englewood Cliffs, 1994.

[GOJ] G. Boolos and R. Jeffrey, *Computability and Logic*, 2nd ed., Cambridge University Press, Cambridge, 1980.

[FRA] N. Francez, *Program Verification*, Addison-Wesley, Reading, 1992.

[GHR] D. Gabbay, C. Hogger, and J. Robinson, eds., *Handbook of Logic in Artificial Intelligence and Logic Programming*, Oxford University Press, Oxford, 1993.

[GAL] J. H. Gallier, *Logic for Computer Science*, John Wiley, New York, 1987.

[GAN] R. S. Garfinkel and G. L. Nemhauser, *Integer Programming*, John Wiley & Sons, New York, 1972.

[HER] G. T. Herman and G. Rozenberg, *Developmental Systems and Languages*, North-Holland, Amsterdam, 1975.

[HOL] G. Holzmann, *Design and Validation of Computer Protocols*, Prentice-Hall, Englewood Cliffs, 1990.

[HOU] J. E. Hopcroft and J. D. Ullman, *Formal Languages and their Relation to Automata*, Addison-Wesley, Reading, 1969.

[HOS] E. Horowitz and S. Sahni, *Algorithms: Design and Analysis*, Computer Science Press, Potomac, Maryland, 1978.

[HU] T. C. Hu, *Integer Programming and Network Flows*, Addison-Wesley, Reading, 1969.

[HUR] M. Huth and M. Ryan, *Logic in Computer Science: Modelling and Reasoning about Systems*, Cambridge University Press, Cambridge, 2000.

[LAW] E. L. Lawler, *Combinatorial Optimization: Networks and Matroids*, Holt, Rinehart, and Winston, New York, 1976.

[MAP1] Z. Manna and A. Pnueli, *The Temporal Logic of Reactive and Concurrent Systems: Specification*, Springer-Verlag, Berlin, 1991.

[MAP2] Z. Manna and A. Pnueli, *The Temporal Logic of Reactive Systems: Safety*, Springer-Verlag, Berlin, 1995.

[MCM] K. L. McMillan, *Symbolic Model Checking*, Kluwer, Amsterdam, 1993.

[MEH] J.-J. Ch. Meyer and W. van der Hoek, *Epistemic Logic for AI and Computer Science*, vol. 41 of Cambridge Tracts in Theoretical Computer Science, Cambridge University Press, Cambridge, 1995.

[MIN] M. Minsky, *Computation: Finite and Infinite Machines*, Prentice-Hall, Englewood Cliffs, 1967.

[PAP] C. H. Papadimitriou, *Computational Complexity*, Addison-Wesley, Reading, 1994.

[ROG] H. Rogers, Jr., *Theory of Recursive Functions and Effective Computability*, McGraw-Hill, New York, 1967.

[SPA] V. Sperschneider and G. Antoniou, *Logic, A Foundation for Computer Science*, Addison-Wesley, Reading, 1991.

[SCH] U. Schoening, *Logik für Informatiker*, B. I. Wissenschaftsverlag, Berlin, 1992.

[TAY] R. G. Taylor, *Models of Computation and Formal Languages*, Oxford University Press, Oxford, 1998.

[TEN] R. D. Tennent, *Semantics of Programming Languages*, Prentice-Hall, Englewood Cliffs, 1991.

[TUR] R. Turner, *Constructive Foundations for Functional Languages*, McGraw-Hill, New York, 1991.

[VDD] D. van Dalen, *Logic and Structure*, 3rd ed., Universitext, Springer-Verlag, Berline, 1989.

[WEI] M. A. Weiss, *Data Structures and Problem Solving Using Java*, Addison-Wesley, Reading, 1998.

Other Useful Mathematical References

[DIE1] J. Dieudonné, *Foundations of Modern Analysis*, Academic Press, New York, 1960.

[DIE2] J. Dieudonné, *Treatise on Analysis*, Academic Press, New York, 1966.

[GOL] R. Goldblatt, *Topoi*, North-Holland, Amsterdam, 1979.

[GRE] M. J. Greenberg, *Euclidean and Noneuclidean Geometries*, 3rd ed., W. H. Freeman, San Francisco, 1993.

[HAR] F. Harary, *Graph Theory*, Addison-Wesley, Reading, 1969.

[HAP] F. Harary and M. Palmer, *Graphical Enumeration*, Academic Press, New York, 1973.

[IYK] S. Iyanaga and Y. Kawada, *Encyclopedic Dictionary of Mathematics*, translated by K. O. May, MIT Press, Cambridge, 1980.

[KRA] S. G. Krantz, *The Elements of Advanced Mathematics*, 2nd ed., CRC Press, Boca Raton, Florida, 1992.

[LAN] E. Landau, *Foundations of Analysis*, New York, 1951.

[LIU] C. L. Liu, *Introduction to Combinatorial Mathematics*, McGraw-Hill, New York, 1968.

[MAC] S. MacLane, *Categories for the Working Mathematician*, 2nd ed., Springer-Verlag, New York, 1998.

[REI] E. M. Reingold, J. Nievergeld, and N. Deo, *Combinatorial Algorithms: Theory and Practice*, Prentice-Hall, Englewood Cliffs, 1977.

[RUD] W. Rudin, *Principles of Mathematical Analysis*, 3rd ed., McGraw-Hill, New York, 1976.

[STR] K. Stromberg, *An Introduction to Classical Real Analysis*, Wadsworth, Belmont, 1981.

[VDW] B. L. van der Waerden, *Modern Algebra*, New York, 1949.

[WEI] E. Weisstein, *CRC Concise Encyclopedia of Mathematics*, CRC Press, Boca Raton, 1999.

[ZAA] A. C. Zaanen, *An Introduction to the Theory of Integration*, North-Holland, Amsterdam, 1965.

Bibliography

[**ADA**] J. F. Adams, On the nonexistence of elements of Hopf invariant one, *Ann. Math.* 72(1960), 20–104.

[**BAR**] J. Barwise, ed.*Handbook of Mathematical Logic*, North-Holland, Amsterdam, 1977.

[**BOO**] W. Boone, The word problem, *Ann. Math.* 70(1959), 207–265.

[**BOT**] I. Borosh and L. B. Treybig, Bounds on positive integral solutions of linear Diophantine equations, *Proc. Am. Math. Soc.* 55(1976), 299–304.

[**BOM**] R. Bott and J. Milnor, On the parallelizability of the spheres, *Bull. Am. Math. Soc.* 64(1958), 87–89.

[**BRIT**] J. L. Britton, The word problem for groups, *Proc. London Math. Soc.* 8(1958), 493–506.

[**BUR**] S. Burr, private communication to M. R. Garey and D. S. Johnson, 1976.

[**BUS**] S. R. Buss, ed., *Handbook of Proof Theory*, Elsevier, Amsterdam, 1998.

[**COH**] P. J. Cohen, *Set Theory and the Continuum Hypothesis*, Benjamin, New York, 1966.

[**COO**] S. A. Cook, The complexity of theorem-proving procedures, *Proceedings of the 3rd Annual ACM Symposium on Theory of Computing*, Association for Computing Machinery, New York, 1971, 151–158.

[**CF**] H. B. Curry and R. Feys, *Combinatory Logic*, North-Holland, Amsterdam, 1974.

[**CUT**] N. Cutland, ed., *Nonstandard Analysis and Its Applications*, Cambridge University Press, Cambridge, 1988.

[DAN] G. B. Dantzig, On the significance of solving linear programming problems with some integer variables, *Econometrica* 28 (1957), 30–44.

[DAV1] M. Davis, *Applied Nonstandard Analysis*, Wiley Interscience, New York, 1977.

[DAV2] M. Davis, *Computability and Unsolvability*, Dover, New York, 1982.

[DLJ] D. Dumitrescu, B. Lazzerini, and L. C. Jain, *Fuzzy Sets and Their Application to Clustering and Training*, CRC Press, Boca Raton, 2000.

[BRI] Encyclopedia Britannica Web article on universal algebra, `http://www.britannica.com/bcom/eb/article /2/0,5716,120652+2+111000,00.html`.

[END] H. Enderton, *A Mathematical Introduction to Logic*, 2nd ed., Academic Press, New York, 2001.

[ELS] S. Even, D. I. Lichtenstein, and Y. Shiloach, Remarks on Zeigler's method for matrix compression, unpublished manuscript, 1977.

[FGJSY] A. S. Fraenkel, M. R. Garey, D. S. Johnson, T. Schaeffer, and Y. Yesha, The complexity of Checkers on an $N \times N$ board—Preliminary report, *Proceedings of the* 19*th Annual Symposium on Foundations of Computer Science*, IEEE Computer Society, Long Beach, 1978, 55–64.

[GHR] D. M. Gabbary, C. J. Hogger, and J. A. Robinson, *Handbook of Logic in Artificial Intelligence and Logic Programming*, vol. 1, Oxford Science Publications, Oxford, 1993.

[GAJ] M. R. Garey and D. S. Johnson, *Computers and Intractability: A Guide to the Theory of NP-Completeness*, W. H. Freeman and Co., San Francisco, 1991.

[GOD] K. Gödel, *The Consistency of the Axiom of Choice and of the Generalized Continuum Hypothesis with the Axioms of Set Theory*, Princeton University Press, Princeton, 1958.

[GRE] M. J. Greenberg, *Euclidean and Noneuclidean Geometries*, 3rd ed., W. H. Freeman, San Francisco, 1993.

[HAL1] P. R. Halmos, *Algebraic Logic*, Chelsea, New York, 1962.

[HAL2] S. Givant and P. R. Halmos, *Logic as Algebra*, Mathematical Association of America, Washington, D.C., 1998.

[HAV] T. Havránek, Statistical quantifiers in observational calculi: an application in GUHA methods, *Theory Decision* 6(1975), 213–230.

[HIG] G. Higman, Subgroups of finitely presented groups, *Proc. R. Soc. of London*, Ser. A. 262(1961), 455–475.

[HOK] J. E. Hopcroft and R. M. Karp, An $n^{5/2}$ algorithm for maximum matchings in bipartite graphs, *SIAM J. Comput.* 2(1973), 225–231.

[HOS] E. Horowitz and S. Sahni, Exact and approximate algorithms for scheduling nonidentical processors, *J. Assoc. Comput. Mach.* 23 (1976), 317–327.

[HRJ] K. Hrbacek and T. Jech, *Introduction to Set Theory*, 3rd ed., Marcel Dekker, New York, 1999.

[HUR] M. Huth and M. Ryan, *Logic in Computer Science: Modelling and Reasoning about Systems*, Cambridge University Press, Cambridge, 2000.

[IYK] S. Iyanaga and Y. Kawada, *Encyclopedic Dictionary of Mathematics*, translated by K. O. May, MIT Press, Cambridge, 1980.

[JEC] T. J. Jech, *The Axiom of Choice*, North-Holland, Amsterdam, 1973.

[JOP] D. S. Johnson and F. P. Preparata, The densest hemisphere problem, *Theor. Comput. Sci.* 6(1978), 93–107.

[JOH] P. T. Johnstone, *Stone Spaces*, Cambridge University Press, Cambridge, 1986.

[JOK] D. B. Kashdan and S. D. Kashdan, Lower bounds for selection in $X + Y$ and other multisets, Report No. 183, Computer Science Department, Pennsylvania State University, University Park, 1976.

[KAR] R. M. Karp, Reducibility among combinatorial problems, in R. E. Miller and J. W. Thatcher, eds., *Complexity of Computer Computations*, Plenum Press, New York, 1972, 85–103.

[KLE] S. C. Kleene, *Introduction to Metamathematics*, Elsevier, New York, 1974.

[KRA] S. G. Krantz, *The Elements of Advanced Mathematics*, 2nd ed., CRC Press, Boca Raton, Florida, 1992.

[LAW] E. L. Lawler, A pseudopolynomial algorithm for sequencing jobs to minimize total tardiness, *Ann. Discrete Math.* 1(1977), 331–342.

[LRKF] J. K. Lenstra, A. H. G. Rinnooy Kan, and M. Florian, Deterministic production planning: algorithms and complexity, unpublished manuscript, 1978.

[LIN] T. Lindstrøm, An invitation to nonstandard analysis, in *Nonstandard Analysis and Its Applications*, N. Cutland, ed., Cambridge University Press, Cambridge, 1988.

[LIP] W. Lipsky, Jr., private communication to M. R. Garey and D. S. Johnson, 1978.

[LIG] P. C. Liu and R. C. Geldmacher, On the deletion of nonplanar edges of a graph, unpublished.

[LOW] P. A. Loeb and M. Wolff, eds., *Nonstandard Analysis for the Working Mathematician*, Kluwer, Dordrecht, 2000.

[LOV] L. Lovasz, Coverings and colorings of hypergraphs, *Proceedings of the 4th Southeastern Conference on Combinatorics, Graph Theory, and Computing*, Utilitas Mathematica Publishing, Winnipeg, 1973, 3–12.

[MAC] S. Mac Lane, *Categories for the Working Mathematician*, 2nd ed., Springer-Verlag, New York, 1971.

[MAA] K. Manders and L. Adleman, NP-complete decision problems for binary quadratics, *J. Comput. System Sci.* 16(1978), 168–184.

[MAS] D. Martin and R. Solovay, Internal Cohen extensions, *Ann. Math. Logic* 2(1970), 143–178.

[MAK] W. J. Masek, Some NP-complete set covering problems, unpublished manuscript, 1978.

[MCA] D. McAllester, `http://www.ai.mit.edu/people/dam/descriptive.html`.

[MEN] E. Mendelson, *Introduction to Mathematical Logic*, 4th ed., Chapman-Hall, London, 1997.

[NEL] E. Nelson, Internal set theory: a new approach to nonstandard analysis, *Bull. Am. Math. Soc.*83(1977), 1165–1198.

[NIS] N. Nissanke, *Introductory Logic and Sets for Computer Scientists*, Addison-Wesley, Reading, 1999.

[NOV] P. Novikov, On the algorithmic unsolvability of the word problem for group theory, *Tr. Mat. Inst. Steklov* 44(1955) *Trans. Mat. Inst. Steklov* (Am. Math. Soc. Translations, Series 2, vol. 9, pp. 1–124).

[PIA] M. Piastra, Probability, modal logic, and fuzzy logic, `http://citeseer.nj.nec.com/piastra98probability.html`.

[PLA1] D. Plaisted, Some polynomial and integer divisibility problems are NP-hard, *Proceedings of the* 17*th Annual Symposium on Foundations of Computer Science*, IEEE Computer Society, Long Beach, 1976, 264–267.

[PLA2] D. Plaisted, Sparse complex polynomials and polynomial reducibility, *J. Comput. Syst. Sci.* 14(1977), 210–221.

[POS] E. Post, Recursive unsolvability of a problem of Thue, *J. Symbolic Logic* 12(1947), 1–11.

[PUS] P. Pudlák and F. N. Springsteel, Complexity in mechanized hypothesis formation, unpublished manuscript, 1975.

[RR1] H. Rubin and J. Rubin, *Equivalents of the Axiom of Choice*, North-Holland, Amsterdam, 1970.

[RR2] H. Rubin and J. Rubin, *Equivalents of the Axiom of Choice, II*, North-Holland, Amsterdam, 1985.

[RUD] W. Rudin, *Principles of Mathematical Analysis*, 3rd ed., McGraw-Hill, New York, 1976.

[SCHA1] T. J. Schaefer, private communication to M. R. Garey and D. S. Johnson, 1974.

[SCHA2] T. J. Schaefer, Complexity of some two-person perfect-information games, *J. Comput. Syst. Sci.* 16(1978), 185–225.

[SCL] C. P. Schnorr and H. W. Lenstra, Jr., A Monte Carlo factoring algorithm with linear storage, *Math. Comput.* 43(1984), 289–311,

[SCH] J. R. Shoenfield, *Mathematical Logic*, Addison-Wesley, Reading, 1967.

[SOL] R. Solovay, A model of set theory in which every set of reals is Lebesgue measurable, *Ann. Math.* 92(1970), 1–56.

[STA] *Stanford Encyclopedia of Philosophy*, `http://plato.stanford.edu/entries/logic-modal/`.

[STE] I. Stewart, `http://www.mcs.le.ac.uk/~istewart/morelAS/BriefDCT.thml`.

[STM] L. J. Stockmeyer and A. R. Meyer, Word problems requiring exponential time, *Proceedings of the* 5*th Annual ACM Symposium on Theory of Computing*, Association for Computing Machinery, New York, 1973, 1–9.

[STR] K. Stromberg, *An Introduction to Classical Real Analysis*, Wadsworth, Belmont, 1981.

[SUP1] P. Suppes, *Introduction to Logic*, Van Nostrand, Princeton, 1962.

[SUP2] P. Suppes, *Axiomatic Set Theory*, Van Nostrand, Princeton, 1972.

[WEI] E. Weisstein, *CRC Concise Encyclopedia of Mathematics*, CRC Press, Boca Raton, 1999.

[WRI] J. D. M. Wright, All operators on a Hilbert space are bounded, *Bull. Am. Math. Soc.* 79(1973), 1247–1250.

Index